EYE ON THE SKY

LICK OBSERVATORY'S FIRST CENTURY

James Lick

EYE ON THE SKY
LICK OBSERVATORY'S FIRST CENTURY

DONALD E. OSTERBROCK

JOHN R. GUSTAFSON

W. J. SHILOH UNRUH

UNIVERSITY OF CALIFORNIA PRESS
Berkeley • Los Angeles • London

University of California Press
Berkeley and Los Angeles, California

University of California Press, Ltd.
London, England

Copyright © 1988 by The Regents of the University of California

Library of Congress Cataloging in Publication Data

Osterbrock, Donald E.
  Eye on the sky.

  Bibliography: p.
  Includes index.
  1. Lick Observatory—History. 2. Astronomical
observatories—California—Hamilton Range, Mount—
History. I. Gustafson, John R. II. Unruh, W. J.
Shiloh. III. Title.
QB82.U62M686   1988   522'.19794'73   87-19118
ISBN 0-520-06109-8 (alk. paper)

Printed in the United States of America
1 2 3 4 5 6 7 8 9

Dedicated to the memory of
Mary Lea Heger Shane

She founded the archives
that preserved the history
of Lick Observatory
and of its part in American astronomy

Contents

Preface

Lick Observatory was one of the first "big-science" institutions in America, built around a large, expensive telescope intended for use by a staff of research scientists, working more or less independently on a variety of problems. It is one of the oldest major observatories still in active operation and producing important research results. The Lick Observatory of 1988 is very different from the Lick Observatory of a century ago. How it started, how it developed, how it changed are what this book is about. Parts of it are about astronomy, but parts of it are about the America we live in. They cannot be separated.

We have concentrated almost entirely upon Lick Observatory. Its history is full and rich, and provides more than enough material for a book. We have left the developments at other observatories and in other countries for other books and other authors. Our aim is to help the reader know Lick Observatory and its history, to understand the observatory and how it became what it is today.

We have organized the book in an approximately chronological sequence, but have concentrated on the great figures of each time—the men who built the observatory in the beginning, and the directors and outstanding astronomers (some of whom were the same!) in the later periods.

Though one of us is a former director of Lick Observatory, one a science writer formerly on the Lick staff, and one a part-time historical lecturer and guide at Mount Hamilton, we must emphasize that this is not an official history of Lick Observatory. We wrote it in our own time, and on our own responsibility. The views that are expressed in it are our own, not those of Lick Observatory, the University of California, or the many participants in the more recent history of the observatory who have allowed us to interview them, or have sent us their thoughts and reminiscences.

This book is intended as a popular history. Much of it, particularly the early part, is based on the books listed in the references for the various chapters. The rest is based on contemporary letters, reports, newspaper and magazine articles, etc. The great bulk of our material came from the Mary Lea Shane Archives of Lick Observatory, located in the Dean E. McHenry Library on the University of California Santa Cruz campus. Mary Shane, a Ph.D. in astronomy and the wife of a Lick Observatory director, was inspired by a visit of Helen Wright to convert the moldering old files of the observatory into a historical archives. It goes back to the beginning of the Lick Trust in 1876, and contains essentially the complete papers of the successive directors since 1888. It includes letters from nearly every well-known American astronomer since Simon Newcomb, and from many European scientists as well. Mary Shane conceived and began to organize the archives on Mount Hamilton, but it came to full fruition after the move of the observatory staff to Santa Cruz. She persuaded Chancellor Dean E. McHenry and Librarian Donald T. Clark to provide space for it in the campus library. She recruited a dedicated band of volunteers who, under her supervision, identified, classified, and filed thousands of letters, clippings, and photographs. She persuaded numerous former Lick staff members and their heirs to contribute additional papers to the archives. The archives was renamed for her in 1982, a year before her death.

Mary Shane dreamed of writing a history of Lick Observatory, but she never got beyond the first chapter. We could not have written this book without the material in the archives she founded, and we have dedicated it to her memory.

We are also very grateful to Dorothy Schaumberg, her successor as head of the Mary Lea Shane Archives, who has helped us immeasurably in tracking down facts, pictures, letters, and clippings dealing with the history of Lick Observatory. She and her assistant Irene H. Osterbrock, the wife of one of us, have been of the greatest assistance in our search for the factual roots of Lick Observatory. We are also grateful to Donald T. Clark and his successors, Librarians David W. Heron and Allan J. Dyson, for supporting the Shane Archives and helping it flourish.

After the Shane Archives, our next most important sources were the University of California Archives and the Bancroft Library in Berkeley. We are particularly grateful to William Roberts, Marie C. Thornton, and Jennifer Prentiss of the Archives, and Irene Moran of the Bancroft for their highly efficient and friendly assistance. In addition we are most thankful to Elizabeth Rentz, for her help in getting material from other libraries, and to Judith Lola Bausch, for her assistance at the Yerkes Observatory Archives. Finally, at the Mount Wilson Observatory Archives, Helen Czaplicki and Joan Gantz, to whom we are also sincerely grateful, assisted us greatly.

Helen Wright generously allowed us to read in manuscript her outstanding new biography, *James Lick's Monument: The Saga of Captain Richard Floyd and the Building of the Lick Observatory*, for which we are most grateful.

We particularly wish to thank the many present and former Lick Observatory students and faculty members who helped us with their memories, either in interviews or by correspondence. For the eyewitness information that they provided we are most grateful to Lawrence H. Aller, Louis Berman, William P. Bidelman, Ernest H. Cherrington, Jr., George H. Herbig, Robert P. Kraft, Gerald E. Kron, Nicholas U. Mayall, Stanislavs Vasilevskis, Merle F. Walker, Harold F. Weaver, and Albert E. Whitford. Likewise we are grateful to Dean E. McHenry and Geoffrey Burbidge for their interviews on the move of Lick Observatory to the Santa Cruz campus. And we owe a special debt of thanks to the late Mary McDonald Wilson who, at the age of nearly ninety-seven, recalled her Mount Hamilton experience which began in 1889 when she was three weeks old.

Finally, we are grateful to Robert P. Kraft, director of Lick Observatory, for permitting two of its Publication Office secretaries to enter parts of this book in the word processor at times when they were free of other more pressing duties. We are especially grateful to those two secretaries, Pat Shand and Gerri McLellan, for their friendly, efficient, and accurate work on this book.

The photograph of Joseph Henry appears with the kind permission of the Smithsonian Institution Archives, for which we are most grateful. The pictures of the Lick House, W. W. Campbell's automobile, and the 36-inch telescope on Christmas Eve are from the private collection of W. J. Shiloh Unruh. The picture of the 120-inch mirror after aluminization, of the three directors (Robert P. Kraft, Albert E. Whitford, and Donald E. Osterbrock), and of the 120-inch Shane telescope and its dome are Lick Observatory photographs. All the rest of the illustrations are from the Mary Lea Shane Archives of the Lick Observatory. We are very grateful to the Observatory and to the Archives for permission to publish the photographs, and to E. Robert Wilson and Diana Larson of the Lick Photolab for their skill and dedication in producing the excellent reproductions for use in this book.

We have incorporated in the book short quotations from numerous letters, nearly all of them from the Mary Lea Shane Archives, the Bancroft Library, or the Library of Congress. Some of the writers were cultivated stylists, most were straightforward scientists, and a few were barely literate misspellers, miscapitalizers, and amateurs in syntax. Nevertheless we have reproduced their words, exactly as they wrote them, in our effort to give "a plain, unvarnished record" of all that happened in Lick Observatory's hundred-year history.

April 13, 1987

1

His Tombstone Is a Telescope
1848–1876

One winter day early in 1848, a rowboat shoved off from the brig *Lady Adams,* anchored off San Francisco. It brought to shore a stern, strong-willed, middle-aged man distinguished by a silk plug hat crowning his tall frame. James Lick had come home to the United States. When he left his Pennsylvania birthplace 30 years before he had only a dollar in his pocket. Now he declared among his belongings a strongbox filled with $30,000 in gold coins, a sight that stunned the customs officer of the poor harbor community. Shrewdly alert to business opportunities, Lick would turn that small fortune into millions in the Gold Rush days soon to come. And his longing for immortality would lead him to finance the construction of a lasting monument, the Lick Observatory, America's first "big-science" research center, which from its beginning set new standards in astronomy.

For setting new standards of eccentricity in burgeoning California, James Lick was without peer. Like many a modern-day Californian, Lick was a transplanted easterner. But his route west took him first through Argentina, and later Chile and Peru. His return to America coincided with the formal ending of the Mexican War, when California officially became part of the United States.

When Lick first surveyed San Francisco, the adobe houses, tents, and ramshackle shanties sparsely dotting the sand dunes and a short stretch of shoreline no doubt dismayed him. The town had barely 1,000 inhabitants. By contrast, his previous home—Lima, Peru—was a metropolis of 100,000 people. Founded by the conquistador Francisco Pizarro in 1535, Lima had been the center of Spanish culture in South America for more than 300 years. In Lima, Lick was welcomed in the homes of the wealthy, the principal customers for his finely crafted pianos. Lick's astute business sense enabled him to capitalize richly on his superb woodworking talents, amassing money with as much skill as he made pianos.

Lick had learned carpentry from his father, John Lick, in Stumpstown, Pennsylvania (later renamed Fredericksburg). John Lick, born Johannes Lück, had Americanized his name on settling in Stumpstown. His father, Wilhelm, had immigrated to Pennsylvania from Westphalia, Germany, in 1765. Wilhelm Lück, who lived to the age of 103, had served in George Washington's army in the Revolutionary War. His tales of the hardships at Valley Forge made a lasting impression on young James.

James Lick was born on August 25, 1796, twenty years after America won its independence. The oldest of seven children, he grew up in a rural setting colored by the cultural heritage of the Pennsylvania Dutch. German was spoken as much as English, and Lick was facile in both languages. At thirteen, he apprenticed in his father's shop, where he learned wood joinery from a demanding perfectionist. His father taught him not only the skills of the trade but also an aesthetic appreciation for the craft. In time, the son mastered the work so well that he met even his father's high standards.

At the age of twenty-one James was still an apprentice to his father, a situation that soon changed as the most dramatic event of his hometown life unfolded. Young James had fallen in love with the miller's daughter, Barbara Snavely, and when she became pregnant he wanted to do the honorable thing. Her father, one of the richest men in Stumpstown, saw Lick's position as an apprentice carpenter as too far below Barbara's social status and refused to allow the marriage. Suitor and miller met in a tense confrontation. The final words of their conversation have been reported so often in biographies of James Lick that they have assumed almost legendary proportions:

"When you have a mill as large and costly as mine, you can have my daughter's hand—but not before!" Snavely roared, dismissing the young apprentice.

"Some day I will own a mill that will make yours look like a pigsty!" Lick hotly retorted as he left.

The miller believed he was setting before Lick an impossible task. Instead, he provided the motivation that propelled the young man from Stumpstown to seek his fortune in the world. His son, John Henry Lick, was born in June 1818, but Lick was not to see him for many years.

Early in 1819, with his father's blessing, Lick left Stumpstown. After a few weeks of searching he landed a job with a piano manufacturer in Baltimore, Maryland, where he worked beside Conrad Meyer, the only close friend of his adulthood. Lick soon decided he could do better on his own, however, and in 1820 he moved to New York City and set up his own shop.

Either he learned how to run a shop quickly or he had innate business skills, for Lick was soon earning a comfortable living in New York. Something he saw, however, caused him to move on before long. Each week, a wagon loaded with furniture, sometimes including pianos he had made, passed by his shop. Curious, he one day followed the wagon to the wharf, where he saw its load put aboard a ship bound for Buenos Aires. Lick reasoned that if the furniture could be shipped to South America and still be sold profitably, he could turn a tidy profit by relocating his business to this seller's market. He laid plans to travel to Buenos Aires, now the capital of Argentina but then part of the newly independent United Provinces of the Rio de la Plata.

Lick invited Conrad Meyer to come along and seek his fortune in a distant land, but his conservative friend preferred the security of steady employment to setting out into the unknown. In those days oversea journeys were long and often hard, and not all ships reached their destinations. And if Lick had known the unstable political climate of his destination he might have chosen to follow Meyer's cautious example. But he was willing to take the risk, as he saw a chance to speedily gain the money he needed to ask again for Barbara Snavely's hand.

In August 1821, Lick began the voyage from New York to Buenos Aires. He quickly found he had no stomach for ocean travel. Sailing ships, tiny compared to today's ocean-going vessels, swayed mightily in rough seas, and Lick was violently ill for his first two weeks at sea. In late November he finally arrived in Buenos Aires. His first years there were difficult. He had to adjust to a new climate and diet, and was frequently ill. He was also at a disadvantage because he did not speak Spanish.

Worse yet, several political factions were battling for control of Argentina. The year before Lick's arrival had been called "The Year of Anarchy" because so many revolutions had assaulted the weak provisional governments. The situation was still explosive when Lick arrived. All the bloodshed he witnessed helped shape his lasting views on human nature.

Most of what is known of James Lick in these days comes from surviving copies of letters he wrote to his father. In one dated March 1, 1825, he shared his view of the strife in Buenos Aires:

"I hope you are in good health and in happy and quiet life, not like me, one minute in the clouds of heaven and the next in the depths of the sea and death always before my eyes in ten thousand forms. This is far from peaceful living and I would not wish it on my worst enemy."

Even so, in the midst of all the turmoil Lick established a prospering business. He quickly learned Spanish, which helped him deal efficiently with his suppliers and customers. And he soon found that the people of Buenos Aires wanted pianos no matter what their political views. Social life in much of South America centered around dances in the home, and a piano was considered the essential centerpiece of a finished household. Those pianos built by Lick, who performed his craft with an artist's eye for detail and perfection, were in high demand. He had as many orders as he could fill. After several years, Lick was prosperous enough to leave his business in the hands of an assistant and take a trip to Europe, where he hoped to regain his health.

He toured England, France, and the Germanic states of northern Europe, including his grandfather's birthplace, Westphalia. Among the sites he visited that made lasting impressions were the famous botanical gardens in London and Paris, the palace at Versailles, and the old structures and ruins that dotted the European landscape. After nearly a year abroad he set sail for Buenos Aires with restored health and fresh ideas for business.

Lick's return voyage to South America was marked by a violent storm that swept one crew member overboard and threatened to sink the ship. It survived only to run into deeper trouble off the Argentinian coast. A fleet of Brazilian ships had blockaded Buenos Aires. They captured Lick's ship, and took crew and passengers across the wide estuary at the mouth of the Rio de la Plata to Montevideo (now the capital of Uruguay). Held in the walled city, Lick wrote that he would risk his life to gain escape. He left his family in suspense for over a year before he wrote again, saying he had arrived safely in Buenos Aires.

Lick had made his escape at the funeral of a fellow prisoner, who had been killed while trying to get away. The funeral was held outside the prison, and Lick managed to hide in the bushes during the confusion following the ceremony. When he reached the border he was feted by a pleased Argentinian officer for having bested his Brazilian captors.

Back in Buenos Aires, Lick got his piano business going full speed again. He also began dealing in the export trade, buying furs and shipping them to England for resale. While vacationing, he had noted the promise of the fur trade; he quickly took his own place in the lucrative market. A few years after his return to Buenos Aires, Lick felt he had earned enough money to visit Stumpstown and persuade Barbara Snavely's father to permit the marriage. The $40,000 worth of furs he brought with him to the United States promised to make the trip a profitable one also.

Lick's homecoming in 1832 should have been triumphant, for he was now a wealthy man. But he learned two things that took any pleasure away. His father had died in 1831, and Barbara Snavely had married. As far as is known, Lick had never written to Barbara during his years abroad, or informed her of his plans to return. She had arranged her life in Stumpstown as best she could, and that did not involve waiting forever for a one-time lover.

Barbara was unwilling to face Lick. She left town and went into hiding as soon as she heard of his return. He spent two futile weeks searching for her and his fourteen-year-old son. Lick had intended to resettle in America but now left filled with ill will toward womankind, a sentiment he held until his dying day. He never again sought the companionship of a wife.

His heartbreak did not affect his will to turn a profit however. After disposing of his furs, he returned to Buenos Aires with a load of staple goods, products in high demand in the war-torn country. The unrest in Buenos Aires was getting worse, compelling Lick to consider moving on. By the end of 1832 he had decided to relocate on the west coast of South America, in Valparaíso, Chile. Here, the uncertainties of historical reports appear clearly. One tradition says Lick traveled over the Andes, not an easy feat even with modern transportation. But a purported letter from him describes a voyage periled by hurricane-force winds through Cape Horn's iceberg-filled waters and treacherous currents.

Whichever route he traveled, once in Chile Lick quickly reestablished his business. He resided in an area outside Valparaíso known for its orchards, and spent his free time in gardening and horticulture. Yet, after only four years, his feeling for the ever-changing political winds in South America caused him to move again, shifting northward and still closer to California. He settled in Peru, which he felt was a more stable country than Chile. On Christmas Day in 1835 Lick arrived at the port of Callao. He located in Lima, nine miles inland, and lived there for the next eleven years.

The Spanish conquistador Pizarro not only founded Lima, the "City of Kings," but also designed its great central plaza and street system, leaving a pattern evident even today. He also laid the cornerstone for its first cathedral, where his remains still lie. From Lima a succession of viceroys ruled over Spain's South American domain. The heart of Spanish power in the New World, the city also became the center of Spain's commercial enterprises. Lima was a bustling, bullish metropolis when Lick arrived to set up shop.

Lick's superb skills earned him entrance to the homes of the wealthy, who would not otherwise have socialized with a mere woodworker. Even so, Lick's income and associations never changed his lifestyle. He remained simple in his tastes and living arrangements, and treated his workers as his equals. To augment his piano trade, he invested in other businesses, including a theatre and a bullfight arena.

During his stay in Lima, Lick began thinking of returning to America. His patriotic feelings, formed while listening to Grandfather Lück's tales of the Revolutionary War, may have been rekindled by newspaper reports of the Texan fight for independence and rumors of a coming war with Mexico. He was certain the United States would prevail and California would soon become American territory. He began making his plans, completing the piano orders he had already accepted and arranging for the sale of his business. It is a measure of his desire to return that Lick sold his business for $30,000, half its appraised value. With his strongbox full of gold Peruvian coins, workbench, tools, and 600 pounds of chocolate, Lick boarded the *Lady Adams*. He arrived in San Francisco on January 7, 1848, less than a month before the Treaty of Guadalupe Hidalgo was signed, officially ending the war with Mexico and ceding California to the United States.

The Mexican War marked the final step in Mexico's loss of control over California, an authority which had been waning for many years. While still a Spanish colony, Mexico's influence had swept northward to the Golden Gate, but had just as quickly receded. A mission and military outpost were established on the San Francisco peninsula the same year that America's war for independence ended, but Mexico's hold on the land was never firm. The territory was largely populated by ranch owners, who worked huge tracts of land granted to them by Mexican officials. Land was also granted to citizens from other countries, as long as they promised to abide by Mexican law.

A steady influx of Americans came to the region, and in time, especially in the northern half of the territory, they dominated the population. It is not surprising, then, that when American forces sailed into the San Francisco Bay in 1846 and claimed authority over the area there was no resistance from the inhabitants.

Under Mexico, the area was known as Yerba Buena, which means "good herb" and referred to a minty plant that grew abundantly on the sandy hills. An official decree in January 1847 changed the name to San Francisco. Military officials also auctioned off property that had been claimed in the name of the government along the peninsula's eastern shore. The auction took place under the authorities' glowing reports of the area's future. Within a year the buyers were feeling disappointed, wondering if they had made a bad choice in their land purchases. They were disillusioned and bitter that the promised growth had not occurred, and had begun to think it never would. When Lick made shore with his strongbox of gold he revived the locals' hopes for the area's future.

Lick's capital attracted attention almost immediately, but he really made a name for himself with his land purchases, which began a month after he arrived. Lick's decision to invest in real estate greatly pleased the residents, who were ready to unload the lots they had purchased from the government. From February through the middle of March 1848, Lick bought 37 lots, at prices ranging from $16 to $3,000. The top-dollar property featured an adobe house with a cellar, where Lick buried his strongbox and stood his heavy workbench over the spot. Five years later Lick sold half of this lot for ten times its purchase price. Today it would command $250,000. Many of the lots that Lick purchased were called "water lots" because at high tide they were mostly underwater. Over the years these lots received a steady supply of fill, and now they form much of the downtown area of San Francisco.

Lick's plunge into real estate was further inspired by the discovery of gold at Sutter's Mill. Many residents, eager to raise a stake and try their hand in the gold fields, sold their property cheaply. As the prospect of gold drew a fresh steam of people through the Golden Gate, San Francisco began to grow and property values to escalate. San Francisco's population increased twentyfold, from 1,000 to 20,000 people, in the two years after Lick's arrival. At the same time, Lick's investments multiplied in value many times over.

The market was wide open for all kinds of investments. Lick wrote a friend of his in Peru, a chocolate maker named Domingo Ghirardelli, that he could do a great business in San Francisco. The 600 pounds of chocolate that Lick arrived with, and subsequently sold at a profit, were made by Ghirardelli. Accepting Lick's advice, Ghirardelli moved north with his family, establishing a company that is still a landmark in San Francisco today.

Lick, too, took advantage of what the market offered. He bought property at the south end of the San Francisco Bay, near the town of Alviso, where he raised produce for sale at the city's markets. Located beside the Guadalupe River, the property was also the site of a mill that Lick rebuilt and expanded until it was one of the handsomest in the land. He invested $200,000 in the mill for hardwood interiors and extensive brass fixtures, proof that money was no object if he could get what he wanted. Area residents, unaware of Lick's boast to the miller Snavely, never quite understood why he constructed a mill as fancy as a ballroom. They often referred to the project as "Lick's Folly."

When the mill was completed in 1855 Lick had it photographed, and sent the pictures to Stumpstown. He could not still have expected to marry Barbara Snavely, and her father was probably dead by this time. He knew nothing of happenings in his hometown since his ill-fated visit, so he must have been acting solely from deep-rooted spite; but Lick had finally made good his angry promise of thirty-five years before.

In 1849, Lick placed his San Francisco properties in the hands of an agent and began concentrating his activities at the mill site in Santa Clara County. He lived in a small, sparsely furnished cabin, rustic accommodations considering his means. In 1860 he built on the property the "Lick Mansion," a twenty-four room, colonial-style house with a marble-framed fireplace in each room. Even after Lick had lived in this spacious home for years, it still lacked drapes and furniture, most of the rooms were unfinished, and its floors were often covered with fruit drying upon newspapers.

Lick built his mill site mansion as a home where he might live with his son, John, and his nephew, James. This nephew had come west to try his luck in the California gold fields and had heard of Lick from his gold-panning cronies. He came to Alviso to see if Lick could be the long-lost relative the family had assumed was dead. It was an unemotional reunion, but Lick was glad to hear about the family and events in Stumpstown, though saddened by the news that his brother had died. At Lick's urging, his nephew stayed for some time and helped care for Lick's properties.

Several years after his nephew appeared, Lick wrote and invited his son to California. John's arrival was greeted by Lick with as little emotion as his cousin's had been. If the dour reticence possessed by son and father had been evenly spread among the locals, the wild San Francisco region would have qualified as a quiet hospital zone.

Despite having similar personalities, the two Licks never got along well. John took little interest in his father's activities, and his "laziness" was a constant irritation to hardworking Lick. John held the title of mill manager, but the old man thought others did most of the work. Perhaps most galling of all, John found the Lick Mansion, which his father had built to draw the family together, entirely too lavish and preferred instead the simple mill site cabin. In this respect father and son were also alike. For all his wealth, James Lick slept on a bare mattress laid across any suitably sturdy support, sometimes a packing crate for a piano, sometimes a door laid across two nail kegs. He never cared for the luxuries his wealth could afford.

Even so, Lick set the hallmark for opulence in San Francisco. In the heart of the city, at a time when the nation was embroiled in the Civil War, he constructed the Lick House, a large three-story hotel marked by elegance. Its main dining room was modeled after one Lick had seen at Versailles. He did much of the interior woodworking himself, especially in the dining hall, which featured fancy crystal chandeliers, stained-glass skylights, intricately inlaid flooring, and artwork by California's leading painters. The pictures, framed in rosewood by Lick's own hands, adorned the dining room's perimeter.

Lick had borrowed money to finance the construction so he would not have to sell any property. Five years after the opening of the Lick House, he had repaid the $400,000 loan. Most of this money came from the $7,500-per-month rental fee he charged for the building. The Lick House stood on Montgomery Street, between Sutter and Post, until its destruction in the 1906 earthquake.

At his property in Alviso, Lick was equally industrious. Devoting himself to horticulture, he developed extensive orchards and filled his gardens with rare plants. He imported many items, including the area's first eucalyptus trees. Uncertain of how well they would grow, he paid to ship several tons of Australian soil across the Pacific as well. Today, judging from the widespread eucalyptus groves in California, it appears that he need not have worried.

Reaching to the East Coast, Lick ordered the framework and glass panes for a huge greenhouse, fashioned after the conservatory he had seen in London's Kew Gardens during his trip many years before. The conservatory arrived, packed in numerous heavy crates. Lick had intended to erect it in San Jose as a gift to the city, but changed his mind after he read an article in that city's newspaper belittling his dress, which typically included his familiar silk plug hat, now a battered headpiece, and a shabby suit. (A woman once took such pity on his appearance that she gave him her husband's hat, not realizing he was probably the wealthiest man around.) The crates remained unpacked until after Lick's death, when a group of San Francisco businessmen, led by Charles F. Crocker, bought the conservatory and erected it in Golden Gate Park, where it stands today as the Conservatory of Flowers.

Lick's own grounds blossomed under his fastidious care, growing to be counted among the most valuable in the state. His devotion to gardening also helped enhance his reputation as an eccentric. Lick would often stop his wagon—a creaking rattletrap that for years looked as if it would surely fall apart in the next mile—to pick up any discarded bones, antlers or hooves he saw along the road. These collections soon grew to large mounds on his property. This behavior appeared odd to the local people, who did not know that Lick milled the bones into powder and used the bone meal to fertilize his fruit trees. Lick's orchards were the most bountiful in the "Valley of Heart's Delight," now known as Silicon Valley.

To tend his property, Lick employed many men who he paid well and worked beside as an equal. He did not hire independent thinkers, however. Lick believed employees should do as they were told without asking questions. To test obedience, he once ordered a gang of new workers to plant trees upside down. Another time, he gave a potential employee the task of stacking, then restacking a pile of bricks. Because the man did the work without questioning his orders, Lick hired him. Thus did Thomas Fraser begin his long association with Lick.

Fraser became Lick's close assistant, after a time rising to the position of foreman at the Lick Homestead, a house Lick had built at the south end of San Jose. He moved there from Alviso in 1870 to avoid the overflow from the Guadalupe River that annually flooded the mill site. During the two-year transfer Lick uprooted and brought along nearly his entire orchard, taking such care that not a fruit tree was lost. Two locust trees were too large to move, however, and Lick cut them off at the ground and had them milled into lumber. He had sections from the very base of the trees sliced into veneer, which he shipped along with some of the lumber to his old friend Conrad Meyer. Lick asked Meyer to use the material to make a piano for the parlor of his San Jose mansion, and gave instructions for placing the veneer so the piano matched a hung pair of doors, twelve feet tall, that led into his parlor.

At the Homestead Lick maintained his simple lifestyle, content to keep his shabby attire and sleep on a mattress spread atop the beautiful piano Meyer had made for him. He lived without a housekeeper, and was lighting the kitchen stove to prepare a meal when he suffered a severe stroke in the spring of 1873. Fraser found his employer sprawled on the kitchen floor, unable to move. With effort, he managed to get Lick into a wagon and to a doctor. Under the doctor's care, Lick soon recovered enough to return to the Homestead. Realizing the burden his weakened condition presented, he shortly moved to a two-room suite at the Lick House in San Francisco, where he might be cared for more easily. Lick had been strong and healthy until then and fiercely independent. His feeble state frustrated him and made him extremely contentious during his remaining days.

After his stroke, seventy-seven-year-old Lick began thinking seriously of the eventual disposal of his estate. He did not want it to go to his son, whom he never legally recognized, his nephew, or other family members. Instead, Lick intended to leave a lasting monument to himself. His first will had provided one million dollars for statues of himself and his parents, huge figures large enough to be seen miles out to sea. He dropped this idea when someone pointed out that such structures would surely be shelled if there were ever a war. Later he conceived the idea of building a giant pyramid, one larger than the great pyramid in Egypt, at Fourth and Market streets in the center of San Francisco. But in the end he was persuaded to build a first-rate observatory.

Supposedly Lick's first contact with astronomy came in the person of George Madeira, a young amateur astronomer and itinerant lecturer, a viable occupation in those innocent, pre-television days. According to Madeira's recollection years after the event, Lick attended one of his lectures in San Jose in 1860. Lick was so interested by what he had heard of astronomy that he invited Madeira to his mansion. Madeira claimed to have stayed there several days as Lick's houseguest, answering questions about the planets and stars, and showing Lick his small, portable telescope. He described the large telescopes of the world to the old man and told of the discoveries made with them. A few years later they met again in San Francisco, Madeira said, and realizing the miser's great wealth he urged Lick to use part of his fortune to build the world's largest telescope.

About the same time as Madeira's second meeting with Lick, Joseph Henry, president of the National Academy of Sciences and secretary (director) of the Smithsonian Institution, visited San Francisco. He stayed at the Lick House, where he arranged to meet James Lick. Henry emphasized to Lick the needs of science, and described how the millionaire could perpetuate his name through a scientific institution, as James Smithson had with the Smithsonian Institution.

Henry's message was reiterated by Louis Agassiz, the famous Harvard naturalist, who gave several lectures in San Francisco at the California Academy of Sciences. His talks, which included stirring appeals for the support of science, were widely reported in the local newspapers. Lick very probably read accounts of those lectures, for soon afterward he made his first recorded gift to science: a valuable lot in downtown San Francisco that he donated to the Academy as the site for a new and enlarged headquarters.

This event put Lick in contact with George Davidson, the man who more than anyone else brought about the gift that led to Lick Observatory. Davidson, then forty-eight years old, was the head of the Pacific branch of the U.S. Coast and Geodetic Survey and a dedicated geodesist and astronomer. Leaving his native England, he had come with his parents to Philadelphia, where he studied under Alexander Bache, who later founded the Coast Survey. In 1850, after joining the Survey, Davidson began mapping the most important harbors on the west coast. Keenly interested in astronomy, he had his own observatory in San Francisco, built around a professional-quality telescope. He was the outstanding scientist on the west coast at the time. And, as president of the California Academy of Sciences, he visited Lick to thank him for the donated property.

Lick and Davidson had met years earlier during court trials to settle the numerous land disputes that arose in the confusion following the Mexican War. Davidson had often testified as an expert witness, and impressed Lick with his knowledge and integrity. Furthermore, Lick liked Davidson. When Davidson called on Lick in his role as Academy president, Lick asked him to return for more conversations. After the old man suffered his stroke, Davidson visited him often in his confinement. They discussed science, astronomy, the planets, the rings of Saturn, and the mountains on the moon. Their talks soon veered to telescopes, and before long Lick decided to forego his pyramid and instead give his fortune for a telescope "superior to and more powerful than any telescope yet made." On October 20, 1873, Davidson announced to an audience at the Academy of Sciences that the generous miser planned to build an observatory in the Sierra Nevada. In his first flush of enthusiasm, Lick authorized Davidson to state that he was willing to spend one million dollars for the project.

By 1874 Lick had set up a Board of Trust to carry out his wishes. The mayor of San Francisco, Thomas Selby, was the board's president and all its members were powerful businessmen and bankers, including Darius O. Mills, a man whose wealth exceeded Lick's and who would come to have a long association with Lick's observatory. Besides the major project of creating the observatory, the trustees were charged to build other public works, both large and small. Gifts of between $25,000 to $60,000 were designated for organizations to aid orphans and the poor; for an institute to train mechanics; for the Society for the Prevention of Cruelty to Animals; and for statues of Wilhelm Lück, Lick's long-dead grandfather, and of Francis Scott Key. Lick had been deeply moved by "The Star-Spangled Banner"; his memorial to the poet can still be found in Golden Gate Park today. Among the larger gifts, $100,000 established a home for elderly widows, $150,000 funded construction of public baths, and $540,000 created a school for mechanical arts, which is still in operation as the Lick-Wilmerding School.

Lick's ideas for bequests often stemmed from specific events in his life. For example, his desire for building public baths is thought to come from the time he got soaked with fish brine and stunk for several days afterwards. It is also clear, however, that several people played their parts in shaping Lick's thoughts, especially with his fixation for attaching his name to the world's greatest observatory. But Lick would not have pursued this grand idea if he did not believe he had invented it on his own. The old man's steadfast independence led him to ignore all advice, good or bad. He could not have been talked into building an observatory, but had to believe he had come upon it himself. Once the idea was fixed in his head, however, he pursued its fulfillment with single-minded devotion. And even though his intention could not be swayed by outside influences, carrying it out involved helpers from many quarters. Their first triumph was in convincing Lick that he should not put the telescope on the property in downtown San Francisco formerly reserved for his grandiose pyramid.

Davidson knew that a telescope in the heart of a city would be useless. The haze, smoke, and turbulent air would drastically limit the science that could be done, no matter how large the telescope. Davidson, who had carried his small telescope all over the west, was well aware how much better the quality of the air is above high mountain sites than in cities, especially those near sea level. He was convinced that the best observatory site would be in the Sierra Nevada, and his arguments steered Lick to the idea of placing the telescope on a mountain.

Few astronomers understood the advantages of mountaintop observing, or had even studied the problem of observatory site selection at that time. Major telescopes had always been erected near a university or town in the lowlands. Building a permanent observatory on top of a mountain was unprecedented, but Lick recognized the importance of doing so. His first choice for a site was a property he owned near Lake Tahoe, but he soon abandoned this spot because he realized it was often snowbound during the winter.

In evaluating potential sites, Lick personally made a trip to Mount St. Helena at the eastern edge of the Napa Valley. This mountain might have become the observatory site but for Lick's unpleasant journey there. He was still bedridden, and undertook the trip lying on a mattress in the back of a wagon. Things went well until the wagon slid on the steep road winding up the mountain, dumping Lick off the mattress and onto the ground. That was enough site surveying for him, and he immediately turned back to San Francisco, vowing not to build his observatory on Mount St. Helena.

In the end, it was Lick's close associate Fraser who came upon the location—Mount Hamilton, located east of San Jose in the Diablo Mountain Range. Named after Laurentine Hamilton, a Unitarian minister who accompanied—and beat—a team of surveyors to the mountain's top, Mount Hamilton satisfied everyone's needs. Over 4,200 feet tall, it had the height that the astronomers deemed essential for a clear view of the heavens. And because it was practically in Lick's backyard (he could see the summit from his property) it pleased the old man no end. Lick decided on Mount Hamilton, provided that Santa Clara County would build a road "first class in every particular" to the summit. Realizing the acclaim such an institution would bring, the county officials quickly agreed, and "Lick Avenue" soon stretched from the valley floor to the mountaintop.

Lick reached this decision in the summer of 1876, just a few months before he died. Yet even with these arrangements in place, Lick was not satisfied. He was worried by the slow progress his trustees were making in converting his property to cash and beginning the public works he had proclaimed. He wondered if his observatory, which interested him more than any of his many other projects, would in fact be completed. To spur forward he opted for a change of trustees, his second such move and one that started tongues wagging over the old man's sanity. Lick brought in a team of doctors who examined him and attested to his mental soundness, thwarting any future challenges to his estate on grounds of his instability.

His final group of trustees was not composed of politicians and bankers. Unlike the influential businessmen who had made up his first two groups, the men on his final board of trustees were largely middle-class professionals, men Lick felt would not get bogged down worrying about legalistic details and with the determination and common sense to get a job done.

Lick's life was slowly fading even as he got the final pieces of machinery in place for executing his will. He grew increasingly weak and listless over the next several months, as the reporters who trooped daily through his hotel room dutifully wrote. At the very end Lick made no sound; his lips moved in silence as he tried, perhaps, to issue final instructions for his trust. A day and a night passed with the doctor believing each breath would be Lick's last. Only his will kept him living, as all strength had drained from his body. With his trustees looking on, Lick died quietly on October 1, 1876: twelve years before his dream of an observatory bearing his name and housing the world's greatest telescope was realized.

2

The Greatest Astronomer of His Time

1874–1878

James Lick had made few friends in his life, but he did not lack for mourners at his funeral. The turnout and fanfare would have been no greater had the wealthy eccentric been a beloved head of state. Lick's body lay in state in the parlor of the Lick House for several days, while the citizens of San Francisco filed past to view his corpse. The Society of California Pioneers, of which Lick was nominally the president (he had accepted the post only under the provision that he would not be required to do anything), handled the funeral arrangements. Lick's casket, elegantly wrought with iron and silver, rolled to the cemetery on a wagon hauled by a team of horses, accompanied by the slow beat of muffled drums. Thousands of people marched in the procession, including California's governor, the mayor of San Francisco, federal and state judges, and nearly all of the area's prominent businessmen. Thousands more paid their respects by lining the funeral route to the Masonic cemetery, where Lick's coffin was entombed until its final ride years later up Mount Hamilton to the Lick Observatory.

With Lick gone, the full burden of carrying out his wishes fell to his trustees, then in their third incarnation. Their predecessors had not been completely idle, even if they had not progressed at a pace fast enough to please Lick. The first trustees, chosen in 1874, lacked any scientific background. For advice on how best to fulfill their charge of building the world's greatest telescope they contacted the man widely considered to be the outstanding astronomer in America, Simon Newcomb, of the Naval Observatory in Washington, D.C. Although by inclination and training Newcomb was a theorist, not an observer, he had worked with telescopes enough to know many important things about them, and about planning an observatory.

Largely self-taught, Newcomb had risen in astronomical circles by impressing those in positions to help him, and proving himself capable when given a chance. He was born in 1835 in Nova Scotia, but through his ancestry considered himself a New Englander. His father worked as a schoolteacher, a post that in those days entailed frequent moves. When the children in one area gained the rudimentary schooling their parents considered necessary, it was time to move on and teach elsewhere. From his father, Newcomb learned early the particulars of words and letters— but not before he learned to handle numbers. Before he was five he could count to one hundred, and soon thereafter busied himself with addition and multiplication. At six and a half he was working on cube roots. Newcomb's talent for mathematics led to his later job as an astronomical "computer," as the people employed to solve lengthy calculations were called.

Newcomb's attraction to astronomy was no doubt also founded in his childhood. His father shared with him a fondness for astronomy and taught him the constellations. Technical books from his grandfather's collection gave Newcomb his introduction to higher mathematics. As he grew older he searched out more advanced works, among them the essential texts used by practicing astronomers. He even labored through Isaac Newton's classic, *Principia Mathematica*. Newcomb would struggle through the complex formulas lacing together the words of these books until he achieved competence with their contents. This understanding opened doors for him later when he met high-ranking scientists.

In the 1850s Newcomb took a teaching job in Maryland, just fifteen miles from Washington, D.C., an easy trek to the city and to the offices of the Smithsonian Institution. On one of his visits to the Smithsonian, Newcomb met Joseph Henry, the institute's secretary (essentially director), who suggested that Newcomb try the U.S. Coast Survey for employment, where he might be put to work mapping the nation's coastline. He had no luck there, but he did become acquainted with Julius E. Hilgard, who was in charge of the Survey. Both Henry and Hilgard were impressed with Newcomb and recommended him for a job with the Nautical Almanac office in Cambridge, Massachusetts, where he began in early 1857.

Working with reams of paper and quivers full of pencils, the Almanac's staff performed thousands of calculations, work now routinely handled by electronic rather than human computers. The Nautical Almanac, today called the Astronomical Almanac, contained numerous tables listing the positions of the planets, moons, and assorted stars for every day of the year, plus the exact times of significant events such as eclipses and occultations. It carried a nautical title and was produced under naval auspices because it had first been compiled to aid sailors. By comparing sextant readings to the book's figures, navigators could accurately determine their location and so find their way across the seas. For astronomers, the Almanac provided—and still provides—important information for planning observations.

While working at the Nautical Almanac offices in Cambridge, Newcomb found time to take courses in astronomy and mathematics at Harvard University and earned his bachelor's degree in a year and a half. He was becoming well established in his field when he received notice in 1861 of a job opening at the Naval Observatory, the national facility that had fulfilled the dream of John Quincy Adams, the sixth president of the United States. In 1825, Adams had delivered an address to Congress in which he urged that a federally funded observatory be established. Lamenting the lack of any such American institution he said, "on the comparatively small territorial surface of Europe there are existing upward of one hundred and thirty of these lighthouses of the sky; while in the whole American hemisphere there is not one." In a later Congressional report, Adams stated his views further, saying, "The express object of an observatory is the increase of knowledge by new discovery. . ., There is no richer field of science opened to . . . the search of knowledge than astronomical observation[.]"

Adams did not achieve his wish as president. After his term ended he entered the House of Representatives, where he continued to lobby for it. His proposals were continually shot down, however, in the complex jockeying for favors that takes place in Washington. Congress refused to appropriate funds for an observatory, but in 1831 it did allow the Navy to set up a "Depot for Charts and Instruments" that housed precision clocks, telescopes, and other astronomical instruments. This meant that for the first time the U.S. Navy had the means to undertake their own navigational work and did not have to rely on other countries for prepared charts.

In 1840 Adams called attention to Russia's new modern observatory at Pulkovo, built by order of the czar. Adams again raised a cry, saying that America was "lagging behind in the race of honor." (President John F. Kennedy evoked similar feelings more than a century later when Russia orbited Sputnik.) Congress responded to Adams' appeals and authorized construction of a building for the Depot, which until then had been operating out of rented quarters. Although popularly referred to as the National Observatory or the Naval Observatory from the time it opened, until 1854 the official designation for the offices remained the Depot for Charts and Instruments. Besides tracking the stars, the staff monitored weather patterns and ocean currents, and provided accurate announcements of the time. The charts and tables they produced were essential for successful navigation, and their work assumed even greater importance during the Civil War, when a naval blockade of the Confederate states played a major part in the Union's war strategy.

The pride of the Depot was a 9.6-inch refracting telescope housed in a dome atop its new headquarters. The building, located near Foggy Bottom on a hill George Washington had reserved for a scientific institution when first laying out the capital city, sat next to a marshy tract on the shores of the Potomac—a river that Newcomb described on his arrival as full of malaria. The site was unhealthy for the astronomers, and was inclement for observing as well. Not only did the moist air make for bad observing conditions but also instruments lost their alignments and their measurements became unreliable as their footings settled in the soft soil.

When Newcomb arrived at the Naval Observatory, he began his career as an observing astronomer. His job consisted chiefly of the classical astronomer's tasks: measuring the positions and times of transit of easily spotted stars when they reached their highest point in the sky. These observations involved instruments with small apertures and finely calibrated scales for determining the exact direction in which they were pointing. The observations yielded refined elements for solving the equations that predict the future positions of the planets. Astronomers collected data at night, and analyzed them the next day, a chore they called "reducing the data," essentially converting a pile of numbers into organized lists of calculations. Newcomb excelled at this task.

Since Newcomb was a theorist by preference and training he had misgivings about joining the Naval Observatory staff. In his autobiography he wrote: "I feared that [t]he drudgery of night work would interfere with carrying on any regular investigation." He also described a practice he and a co-observer followed that illuminates his dedication to the work: "When either [of us] got tired, we would 'vote it cloudy' and go out for a plate of oysters at a neighboring restaurant." Despite not being whole heartedly devoted to observing, Newcomb did learn some important points about organizing and operating an observatory. He quickly realized that for any good work to come from an observatory it must first possess high-quality equipment. Furthermore, the people operating the equipment must also be of the highest caliber. In its early days, the Naval Observatory fell short of its European counterparts in both these criteria.

The observatory's superintendent ordered new equipment in an attempt to come up to the mark, and Newcomb was assigned the task of going to Europe and studying the methods employed there. He was impressed by the uniform approach European astronomers used in reducing their observations, as opposed to the Naval Observatory's style of every man for himself. One of Newcomb's lasting contributions to American astronomy was to convince his colleagues of the importance of standardizing their data reduction, thus enabling them to match their European competitors in quality and reliability of work. Visiting Europe gave Newcomb the opportunity to meet with some of the best astronomers and instrument makers of the time, and he made good use of these acquaintances in later years.

Newcomb took on his greatest responsibility at the Naval Observatory when he assumed charge of the 26-inch telescope, the largest operating refractor in the world in 1873 and, at the time, the crowning achievement of the optical firm of Alvan Clark and Sons. The Clarks were regarded as the finest technicians (some said artists) in the United States (if not the world) for producing the precise curves on glass disks to make them effective lenses. The Clarks built identical 26-inch lenses for the Naval Observatory and for Leander J. McCormick, who had made his money in farm equipment and wanted to use some of it to buy a telescope for the University of Virginia. McCormick had placed his order first, and when the Naval Observatory requested a still bigger lens the Clarks refused, saying that to do so would dishonor McCormick. They agreed, however, to make another of equal size.

Unfortunately for McCormick, although he had a telescope it took ten years to convince the University of Virginia to take possession of it and put up funds to operate it. The Naval Observatory, in contrast, had the backing of the federal government, and had its telescope up and running soon after the Clarks completed it. The Naval Observatory's large refractor, however, did not begin to live up to its capacity for scientific achievement until Newcomb returned to theoretical work and the telescope was assigned to the more observationally oriented Asaph Hall in 1875. But as the astronomer in charge of it, Newcomb was the natural person for Lick's first Board of Trustees to approach in 1874 for advice as they started on their great project. Trustee Darius O. Mills came to Washington and met with Newcomb several times during the summer and fall of 1874. Newcomb invited his young assistant, Edward S. Holden, to take part in these meetings.

Holden became increasingly involved in the project through his energy, strong interest, and persistence, and soon began to influence the development of Lick Observatory. The twenty-eight-year-old astronomer had come to the Naval Observatory in 1873 and had first been assigned to work on the meridian circle, a special telescope used to measure the time a star crosses the meridian and its altitude above the horizon when it crosses. Holden joined Newcomb on the 26-inch refractor when it came on line.

Holden had been born in St. Louis, Missouri, in 1846. His mother died when he was three, and after that he was raised by an aunt in Cambridge, Massachusetts. He received a private school education while growing up, and his schooling was enhanced greatly by the many scientists and intellectuals who lived in the Cambridge area. Some of them frequented his aunt's household. His first introduction to astronomy came when his cousin, head of the Harvard College Observatory, showed him the stars through a 15-inch telescope. When he was sixteen, Holden entered a private high school in St. Louis; after graduation he enrolled at Washington University in the same city. In 1866, just before his twentieth birthday, he earned his bachelor's degree in mathematics and astronomy. Fortunately, Missouri was stable enough for Holden to complete his studies while the Civil War was underway.

Holden had the good fortune to study under William Chauvenet, a mathematics and astronomy professor who was also chancellor of the university. Not only did he receive a solid education from his mentor, but he also established a relationship with a respected and widely known figure in American science, which in turn increased his credibility among the scientific community. Through Chauvenet, Holden also met his future wife, Mary Chauvenet, his professor's daughter.

After receiving his degree at Washington University, Holden applied for and received an appointment to West Point. His prior education made him one of the top students in the corps of cadets, and his instructors were quickly impressed with his abilities. In 1870 he graduated from West Point third in a class of fifty-nine students, received his commission as a second lieutenant, and spent a year with a field artillery unit. He then returned to West Point as an instructor, serving up the traditional style of learn-by-rote education at which he had excelled as a student.

Holden had great abilities to assimilate and organize information, but he never demonstrated much talent for creative thinking or inspired research. While still a West Point instructor he published his first technical paper, an observation of the spectrum of the earth's aurora, but his work was too inexact to be of much scientific value. He was never formally trained in the methods of scientific research, and never learned it fully in his later career despite rubbing elbows with some of the most creative scientists of his day.

Holden had known for years, even before he studied with Chauvenet, that he wanted to work in astronomy, and from West Point he applied for a job at the Naval Observatory. He was offered a position there as professor, and promptly resigned his army commission and moved to Washington, D.C. From his initial assignment on the meridian circle, it was a considerable step up when he became Newcomb's assistant, operating the 26-inch refractor. Holden must have impressed Newcomb, because the older astronomer did not hesitate to invite his young assistant to take part in the meetings with Mills when the Lick trustee came to ask Newcomb's advice.

One of the first issues Mills raised with Newcomb was whether any firm besides Alvan Clark and Sons was competent to build the telescope. James Lick, on advice from Joseph Henry, had written Alvan Clark earlier and asked how large a telescope it was possible to build and what its cost would be. The elderly optician could not easily answer. Making a telescope depended first on obtaining glass disks to shape into the lens, and glassmakers had already strained their talents providing disks for the largest existing telescopes; how much further they could push their craft was not known. Secondly, the Clarks had no a priori knowledge of the limits to lens shaping. For each successively larger lens they made, they scaled up their design a little more, but until they had completed one and tested it they could not say what the prospects would be for the next larger one. Not until after the two 26-inch telescopes were completed did Alvan Clark finally announce his conclusions to Lick, quoting a price of $180,000 in gold for an instrument of "36 inches clear aperture."

The figure even shocked wealthy James Lick, and prompted Mills' question to Newcomb of other likely vendors. The Clarks were the best in America, but it was possible that lensmakers of comparable talent were operating in Europe. To answer the question Newcomb departed for Europe in December 1874 as an agent for the Lick Trust, leaving Holden in charge of the 26-inch in his absence. He traveled to England, France, and Germany, and met those countries' top astronomers and opticians. In addition to asking them about manufacturers, he also sought their advice on what telescope design would best satisfy Lick's goal.

Even though all the important telescopes at that time were refractors, it was not obvious that the trustees should also adopt that design. Refractors, resembling oversized spyglasses, focus light with lenses. Light is bent, or refracted, as it passes through the glass; the symmetric shape of a convex lens causes the light transmitted to bend inwards as a converging cone, creating a concentrated image at the focus.

To image stars cleanly, however, requires a very clear lens. The glass must be free of bubbles or internal blemishes, which would block the light's passage, and also of stresses, which would alter the refractive properties of the glass, causing an imperfectly converging light cone and a poor image. Even with a perfect lens, however, images focused by refraction cannot escape some distortion because each wavelength or color bends at a slightly different angle. (Prisms, diamonds, and cut glass make use of this effect; rainbows result from it.) Because of this refraction effect, astronomers combine two lenses of different shapes and types of glass to form a telescope's single objective lens; the pair compensate for one another, and minimize the prismatic smearing.

Focusing light with a concave glass mirror rather than a convex lens, as is done with all large telescopes today, avoids this chromatic problem completely because all colors reflect at the same angle and converge to the same point. Also, because the top or front surface forms the mirror, the internal qualities of the glass disk are largely unimportant—the light does not penetrate into it. The glass acts only as the material that is figured into the exact shape necessary to focus the light; its surface, coated with a layer of reflective metal, is the actual mirror.

Unfortunately, in Newcomb's day the technology for making reflecting telescopes was far less developed than that for producing refractors. Astronomers had not learned to make large glass mirrors, coat them with a microscopically thin layer of vacuum-deposited aluminum for high reflectivity, and mount them to maintain their ideal paraboloidal forms and concentrate all the light from a star into a bright, point-like image.

Some of the earliest reflectors, in fact, had been made from metallic disks polished to a high sheen. Large disks of "speculum," a mixture of tin and copper, were easier to obtain than glass ones, but were difficult to shape accurately because the soft metal deformed so easily. Metal mirrors also tarnished quickly and required constant polishing. Because of the problems with reflectors, most professional astronomers of the nineteenth century regarded refractors as their first choice in telescopes.

Still, a farsighted review of the advantages of reflectors was sent to the Lick Trust by Scottish astronomer David Gill, whom Newcomb met during his travels through Europe. Gill astutely recognized that the classical work of astronomy—accurate measurements of star positions— would soon be superseded at the frontier of science by astrophysics: interpreting objects and events in the heavens in light of physical knowledge gained in the laboratory. Astrophysics makes up the bulk of astronomical work today. With this new science in mind, Gill wrote that the telescope chosen should be a reflector because it would be free from chromatic distortion, could be made with a larger aperture and so gather more light from faint objects, and in a "fast" optical design, with a small focal ratio (or small f-number for camera buffs) to allow shorter exposure times.

Newcomb too, for a time, favored a reflector, but in the end wavered in his choice. "We know pretty well the utmost that can be done with the refractor, but do not know what might be done with the reflector," he wrote to the Lick trustees. And in a letter to England's George Calver, who was pushing hard for a reflector, Newcomb said, "In order to reach a conclusion in favor of the reflector I would have to make a more complete examination of the performance of such instruments than I am now able to do. At the same time I do not deny a certain probability that the great telescope of the future will be a reflector." As a realist in the present, however, when Newcomb toured Europe he sought design ideas for a refractor.

Newcomb submitted a detailed report of his European findings to the Lick trustees in March 1875. He was impressed by two firms that could provide the raw glass disks to be made into the lenses. One, Chance & Company of Birmingham, England, had cast the blanks for the Naval Observatory's large refractor. The other, Charles Feil in Paris, France, struck Newcomb as offering exceptionally high-quality work. Shortly after his visit Feil demonstrated his talent for astronomical ware by casting the 30-inch blanks for Pulkovo Observatory's new refracting telescope. As for someone to shape the glass into lenses, Newcomb brought back no word of anyone whose skills matched those of Alvan Clark and Sons. He did, however, later receive a strong recommendation from Gill that Howard Grubb, of Ireland, should be hired to make Lick's telescope. Gill wrote, "as a designer, as a man of science, and as a man of internal resources, energy and skill Grubb is unparalleled." Grubb himself was excited by the project, and envisioned at various times reflectors of 6-foot diameter or greater that he would make of annealed metal for the Mount Hamilton site, but he was equally thrilled by the chance to make a large refractor for the trustees.

To go with Newcomb's report from Europe, the trustees had in hand preliminary plans for an observatory from the original meetings between Mills, Newcomb, and Holden. Mills had verbally outlined what Lick wanted to accomplish—to build the world's largest telescope and provide a suitable observatory to support its use in front-rank scientific discovery—and Newcomb and Holden had described in general terms how this might be carried out. Holden, working from notes he had taken at these meetings, had written up the ideas discussed, filled in the empty spaces surrounding Newcomb's general descriptions, and sent a detailed report in October 1874. This report marked Holden's formal appearance as an advisor to the Lick Trust, and came about because Newcomb himself was extremely busy with his research and with responding to requests for advice. (Just as the Lick trustees had sought him out, others approached Newcomb for his high-powered opinions.)

Newcomb was glad to hand over some of his responsibilities to his ambitious young assistant.

Holden's report included some rough plans for the observatory buildings. It showed a long, low building that would contain offices, computing rooms, and library space. A dome rose at each end of the building, a small one to hold a modest-sized, easily acquired telescope, and a large one for Lick's great refractor. The building was planned to have an open hallway running its entire length for possible use in optical experiments. Surrounding this Main Building were smaller structures to hold the auxiliary observing instruments: the meridian circle for determining star positions, a transit telescope for fixing precise times, and a heliograph for studying the sun. The report also listed the tools and equipment a modern observatory should possess, described how the administration should be organized (influenced, no doubt, by the military bureaucracy under which Holden and Newcomb worked), and suggested what areas of research should be pursued. Newcomb and Holden also agreed that the Mount Hamilton site should be evaluated by an experienced observer before the trustees considered the location as final, a proposal Alvan Clark had also made in his communication with Lick nearly a year earlier.

Newcomb further stressed to Mills the need to choose early a director for the observatory, someone astronomically trained who could maintain supervision during the observatory's construction. At the Naval Observatory, Newcomb worked under a career navy officer, who could provide little more than bureaucratic expertise; the astronomers often had to explain the aims of their research to a leader who lacked the background to appreciate them. Usually, the superintendent respected the astronomers' judgement enough to support their intended line of research without understanding deeply any of their goals. Holden, with his solid West Point background, had no difficulty operating within a military chain of command, though Newcomb sometimes thought it inefficient.

Newcomb was very likely offered the Lick directorship at this time, but turned it down. He probably saw such a position as yet another intrusion on his desire to return to theoretical research. And the prospect of life on an isolated mountaintop, with a wife and two young daughters to care for, held no appeal for him at the time. Newcomb promptly recommended Holden for the job, even though his young assistant had been working as a professional astronomer for less than a year. Whether Holden was offered the position at this time is not clear, but he continued to pour so much advice on the trustees that they could not ignore him, and eventually offered him the post. From 1874 on, Holden made career decisions under the assumption that he would one day be in charge of Lick Observatory, and that anything else was but a step toward that post.

If Newcomb had then accepted the directorship he probably would have greatly regretted his decision, for following his European trip several events slackened his enthusiasm for the project. Shortly after he delivered his report on the glass makers and opticians, the first group of trustees resigned under pressure from Lick, who felt they were moving too slowly in carrying out their task. Lick also dissolved his original Deed of Trust, for he realized that in it he had signed away any rights to change either the personnel of his trust or the wording of his trust deed. Hearing of these difficulties, Newcomb wrote Mills to say, "The present discouraging prospect of the Lick Observatory, naturally diminshes the satisfaction with which I should have looked upon [my part in] the execution of that noble project."

Newcomb's disappointment did not compare with that experienced by George Davidson, an earlier scientific advisor to James Lick. Davidson, president of the California Academy of Sciences, was one of the men who had helped turn Lick to the idea of donating his money to build a telescope, and was in fact Lick's original scientific advisor. He consulted with eastern astronomers for Lick, and provided the old man with an estimate of $1,500,000 as the likely cost for the astronomical monument he had in mind. Lick considered this figure too high, and wanted to provide only $500,000 for the job. After long discussions with the old man, Davidson thought he had convinced Lick to raise the amount to $1,200,000. The Academy president also believed he had gotten Lick to agree to build the observatory in the Sierra Nevada Range at an elevation above 10,000 feet, where he believed the skies would be superb for observing. Davidson embarked on a trip to Washington, D.C., confident that he had shaped the project's direction. In advance of his fact-finding trip he wrote to Newcomb saying, "The observatory proposed by Mr. Lick will be built, and I can tell you confidentially that over a million in gold will be assigned to it. Mr. Lick is a man of strong views and some of them might be called eccentric, so that in certain matters he may be very stubborn, but I have faith that all will end rightly."

When Davidson returned to San Francisco, however, he learned that stubborn and eccentric Mr. Lick had executed his first Deed of Trust, providing only $700,000 for the telescope and observatory. In Davidson's absence, members of the Society of California Pioneers undoubtedly had a chance to discuss with Lick the disposition of his fortune. They, along with the California Academy of Sciences, were due to receive whatever was not spent in fulfilling Lick's bequests, so any money not designated for the observatory would be more likely to come to their coffers. Davidson was further infuriated when he learned that Lick now specified his observatory be placed not on the high Sierras but on some property he owned beside Lake Tahoe. Later, when Lick again changed the observatory site and settled on Mount Hamilton, Davidson refused to have further dealings with the old man or his project. The premier West Coast scientist now stood on the sidelines. Although disgruntled by the turn of events, Davidson remained an interested observer of the trust's execution as a member of the California Academy of Sciences, which had been promised a financial gift in the Deed of Trust.

Fortunately for the Lick trustees, Newcomb did not feel as great a personal involvement in the observatory nor was he as easily discouraged as Davidson, or they might have lost his participation as well. After dissolving in his first Deed of Trust, Lick did not establish a new one or select a new Board of Trustees for several months, leaving the project in limbo. Then, near the end of summer 1875, he named Richard S. Floyd as president of the new trust, and also appointed his own son, John, as a trustee. Even though he refused to recognize his son legally, he finally consented to leave John $150,000 in his will. Lick's first trustees had argued that by leaving his son entirely out of the will he was opening the way for John to challenge the estate. Lick refused to listen to them, but apparently was willing to heed their advice after he had removed them from office.

The new Deed of Trust also contained a provision that the telescope and observatory should be conveyed "to the corporation known as the Regents of the University of California" on completion. John B. Felton, Lick's attorney and a University of California regent, encouraged his client to hand the observatory over to the university. Additional support for this plan most likely came from Joseph Henry, who headed the Smithsonian Institution. Henry probably first suggested the idea in one of his letters to Lick. When McCormick had ordered his large refractor from the Clarks, Henry had written him, detailing his ideas on observatories and directors. In this letter he stated that the director should be associated with an important college or university to add to the observatory's reputation. No doubt when Henry wrote Lick he gave him similar advice.

The year before Mills met with Newcomb, Henry had also replied to a letter from Lick with considerable advice on how to build and equip his observatory. Henry, a physicist, was convinced that Lick should create an observatory devoted to astrophysics, and that it should be run by an astronomer well trained in physics rather than in the conventional brand of astronomy. Henry even went so far as to suggest that Lick hire J. Norman Lockyer, a famous British astrophysicist, to head his observatory project. Henry advised Lick not to try for the biggest telescope possible, for "if you wait until [it] can be finished, your life will probably be terminated before any results can be produced." Instead, Henry urged Lick to aim for a well-instrumented astrophysical observatory, which could begin doing important science immediately.

But Lick rejected Henry's suggestions, and clung fast to his goal of building an observatory that would house the world's largest telescope. In doing so, he proved Henry a prophet, for he died before his telescope could be ordered. Even Lick's strong will was not enough to push the project to completion. That required the united efforts of two determined individuals: Floyd, the new trust president, and Thomas Fraser, Lick's faithful employee. The task almost proved beyond their combined strengths as well.

3

A Son of the Confederacy

1876–1884

The Diablo Mountains stretch southward from the east side of the San Francisco Bay for over 200 miles. The material forming the mountains, mostly shale and sandstone that was once ocean bottom, has been sheared and shattered by the large-scale tectonic movements shaping the region. As the westward-moving North American plate slides over the Pacific plate, it pushes under and compresses the ocean floor. About a million years ago, a ridge of this subducted material was squeezed and uplifted to form the Diablo range, land left high and dry, paralleled and crossed by earthquake faults.

The former sea floor that makes up the Diablos is largely devoid of fossils, so geologists believe the material came from a cold, deep-water zone not conducive to ocean life. But in the million or so years following its uplift, the mountain range acquired a full topping of oak and madrone trees, with plentiful game, such as deer, mountain lions, and bears, slipping through the dense underbrush. In 1861, on the first recorded trip to the range's highest point, surveyors scouting the mountain for the State Geological Survey were forced to leave their horses and continue on foot when the brush became too thick. It took them all day and part of the night to get from their camp at the base of the hills to the summit and back. On their trek, the surveyors were accompanied by Laurentine Hamilton, a Unitarian minister who preached in the San Jose and Oakland areas. Hamilton did not have to lug heavy surveying equipment and was able to race ahead of his companions as they neared the summit. From above them, he waved his hat and shouted, "First on top, for this is the highest point." In their report on the mountain, the surveyors suggested that it be called Mount Hamilton in honor of the minister who led the final ascent.

Over 4,200 feet tall, Mount Hamilton tops even Mount Diablo, the namesake for the range. James Lick had fixed in mind an elevation of at least 4,000 feet for his observatory. When Thomas Fraser, Lick's trusted assistant, examined the survey records in 1875 he found that Mount Hamilton met this criterion. He made a trip to the mountaintop to judge its suitability and was impressed by the clear skies and stunning views. To the west and north lay the Santa Clara Valley and the southern end of San Francisco Bay. Beyond the coastal mountains to the southwest sparkled the Monterey Bay, and far across the San Joaquin Valley to the east rose the high, snow-capped peaks of the Sierra Nevada. Fraser reported favorably on Mount Hamilton to his employer, but as Lick Trust President Richard S. Floyd later wrote, "[T]he possibility that a complete astronomical establishment might one day be planted on its summit seemed more like a fairy tale than a fact. It was at that time a wilderness. A few cattle ranches occupied the valleys around it. Its slopes were covered with chaparral or thickets of scrub oak. Not even a trail led over it."

Nevertheless, Fraser's report ended the long search for an observatory site that had led from the Sierra Nevada to Lake Tahoe to Mount St. Helena and, finally, to a spot in Lick's backyard. The old miser accepted Mount Hamilton, and Santa Clara County agreed to construct a road to the summit. In late 1875, County Surveyor Andrew T. Herrmann laid out the route for the 26-mile road over the hilly contours. He had to follow a winding path to insure that the road's slope never exceeded a six-and-a-half-percent grade, a limit imposed by the abilities of mules and horses to haul heavy wagon loads uphill. The road today still follows this sinuous route, to the great displeasure of those who easily become carsick.

In February 1876, the county awarded a contract for building the road. A work crew of 150 men began the job, and the force swelled to over 600 by summer, as the workers blasted, graded, and spread gravel in a race to finish the job before the winter rains came. It was a tie: "Lick Avenue" was completed in December 1876 at a total cost to the county of over $70,000. With a first-class road in place it was time to begin preparing the summit for the observatory's construction. The task of bringing an observatory to the isolated and rugged mountain peak fell to Floyd and Fraser, and was perhaps the most significant experience in their lives.

Floyd was born in Georgia in 1843, a year when Lick was still busy making pianos in Peru. Floyd's parents died while he was young and he was raised by his grandparents, a child in a privileged family of the southern aristocracy. His grandfather operated a shipyard where Floyd spent much time as a child. He learned to maneuver small boats along the inland streams and coastal waters, activities that filled him with a longing for the open ocean. In his early schooling he displayed talents for mathematics, writing, and drawing. At the age of sixteen, he was appointed to the United States Naval Academy in Annapolis, Maryland, and received a formal education for the first time.

As a midshipman, Floyd stood out in many of his classes and quickly gained the admiration of his professors. Although he disliked the rigor of the military academy—a far cry from his leisurely existence on the family plantation—he thrilled at the chance of a real ocean voyage on the midshipman's training sloop. During his student years tensions between northern and southern states were growing, often erupting into loud arguments between midshipmen divided along regional lines. When Georgia seceded, Floyd knew he had to decide whether to remain loyal to his home state or side with the federal government. In 1861, when Fort Sumter was fired on and the Civil War began in earnest, circumstances forced Floyd to choose. He resigned his commission at the Academy, as did many of his southern classmates, traveled south, and enlisted in the Confederate States Navy.

While waiting for his ship to be outfitted and sent to sea, Floyd heard of the destruction of his family's plantation, which undoubtedly fired him to seek revenge. Still, as a naval officer, he pursued his revenge under the gentlemanly rules of war that nations followed in those days. Floyd served as second officer on the C.S.S. Florida, a "sea-raider," or fast Confederate gunboat, that preyed on tradeships heading to and from northern ports. To join the action, the Confederate ships had to skirt a tight Union blockade around the major southern ports. Ironically, Simon Newcomb's work at the Naval Observatory was helping make this blockade more effective and, consequently, Floyd's life more difficult.

Once his ship set out to sea, a typical operation involved sighting a heavily loaded freighter, giving chase, and offering the unarmed vessel the choice of surrendering or being blasted and boarded. All the ships they overtook wisely surrendered. The Confederate ship would take on any valuable cargo to help finance the war effort. Crew and passengers, along with their personal property, would be delivered to a neutral ship or port. The captured ship would be burned or confiscated if it was of value to the Confederate navy. Floyd's naval ventures took him from South America to New England in search of prey. As needed, his ship would head for England or France for refurbishing, shore-leave for the crew, and a rendezvous with Confederate agents for pay.

At one point the Confederate sea-raider was anchored in Salvador, a

Brazilian harbor, at the same time the *U.S.S. Wachusett* was laying over there. After many successful sea raids, Floyd's ship was well known to the Union navy, including the captain of the *Wachusett*, and this reputation proved the ship's undoing. The Confederates fully expected to battle their foe on the open sea, but the Union captain looked on the anchored vessel as an irresistible plum, ripe for picking. Violating the port's neutrality and opening an international incident, the *Wachusett* rammed the *Florida* under cover of darkness, disabling the ship and forcing the crew to surrender. Floyd, his fellow officers, and the rest of the crew were taken prisoner and transported to Massachusetts. After several months in a military prison, they were released in February 1865 after signing a parole agreement to leave the United States. Floyd managed to secure passage to Europe and was still trying for an assignment to another vessel when the Civil War ended.

With the war over and prospects in the south bleak, Floyd headed for Central America and traveled overland through Panama to the Pacific. Working his passage on a steamer, he arrived in San Francisco Bay the day after Christmas in 1865, a twenty-two-year-old ex-Confederate naval officer recovering from the South's defeat and looking for a fresh start. He first took a job as a fireman shoveling coal on a steam ship—a post far below his training and status—for a line that plied between Mexico and San Francisco. This hard work did not sit well with him, and he next tried a borax mine in northern California's Lake County, where he found that many skills he had learned aboard ship were invaluable. (It is some measure of fate that he traveled there with John Fraser, the brother of Lick's assistant, Tom.) This was Floyd's introduction to the region around Clear Lake, an area he soon adored and later made his home.

After a few years at the borax mine, Floyd foresaw a dead-end to business opportunities there and returned to San Francisco. He was hired in 1868 for a job more worthy of his talents, captaining a ship that sailed between San Francisco and Hawaii. On a return trip from the islands several years later, Floyd met Cora Lyons and her father, who were passengers on the cruise. Floyd was immediately taken with Cora, and she with him. Following a brief romance they were married in San Francisco in September 1871, and honeymooned in the Clear Lake region. The society pages in the San Francisco papers raved about the handsome young couple.

Floyd found in Cora's life many similarities to his own. She, too, was from the south, had lost her mother at an early age, and had seen her home destroyed in the Civil War. She had come to San Francisco to be with her father, a retired judge who had salvaged considerable funds from his southern holdings. In California, he had invested and reinvested his money wisely, and had become a very wealthy man. Judge Lyons died a few years after his daughter's marriage, leaving her a huge inheritance. This, plus Cora's misgivings about her husband's sometimes dangerous life on the ocean, prompted Floyd to settle down at the ripe old age of thirty-one. In 1874, the Floyds bought some 300 acres along the shore of Clear Lake, where they built their family home, called Kono Tayee (an Indian phrase meaning Mountain Point). They divided their time between this estate and a magnificent townhouse they maintained in San Francisco, where they indulged in the gaiety of high-society living.

Floyd's path first crossed Lick's while the old man was considering sites for his observatory. Lick thought of Mount St. Helena, which is very near Clear Lake, as a possible location. Fraser, who knew of Floyd through his brother as well as through write-ups in the local papers, suggested to Lick that Floyd could provide valuable advice on the region. When the two met, Lick was immediately taken with Floyd's southern congeniality and wide experiences, which made him a captivating conversationalist. Apparently, Floyd's self-assurance, an aura of command gained during years at the helm, also impressed Lick. On the basis of this one meeting he selected Floyd to head his second Board of Trust. He saw in Floyd the technical expertise and personal determination he deemed necessary to carry out his observatory project.

While Floyd supplied the expert guidance to build the observatory and telescope, Fraser provided the drive and on-site inventiveness necessary to complete the project. Although they were equal in their dedication to the task, Fraser always looked to "the Captain" for leadership. He once wrote to Floyd: "Let me know your wishes in the matter and they will be obeyed to the letter . . . You will always find me as I have always been, to attend to your instructions as near as mortal can." Fraser, born in Nova Scotia, had arrived in California in 1866 at the age of twenty-four. He was a skilled cabinetmaker and, as shown by his first job stacking (and restacking) bricks for Lick, he could follow orders to the letter. Both qualities no doubt helped earn Fraser a place with Lick, who hired him in 1873. Within a short time, Lick was relying increasingly on him to manage the Homestead property and handle other business matters. Based on Fraser's loyalty to Lick and his sound knowledge of mechanics and construction, the trustees appointed him as Superintendent of Construction for Lick Observatory.

In 1876, while the road to Mount Hamilton was nearing completion, Fraser took a horseback trip to the mountaintop to view the progress. Reaching the summit, he found a surprising sight: three squatters had set up a shack there. At the time, the property still belonged to the federal government, and squatters' rights held. If they could occupy the land continuously for a set time, they could claim title to the land. No doubt they planned to do just this, and then sell the land to the Lick Trust for a considerable sum. Lick had had similar problems with vagrants when he was accumulating lots in San Francisco, and his solution then was to hire some strongarms to throw them off his property. Fraser followed suit, after getting permission from Floyd for this course of action. Backed by hired men armed with rifles, Fraser chased the squatters away, then set fire to their shack. He and his crew erected a crude cabin, and Fraser paid his men $2.00 a day to stay there, maintaining a continuous presence to thwart the squatters' return. The issue of ownership was settled in June 1876 when an Act of Congress transferred 1,350 acres atop Mount Hamilton to the Lick Trust. Other grants and purchases later added to the trust's summit property.

The "Battle for Mount Hamilton" occurred just before Floyd departed on a yearlong journey to prepare for building the observatory. In actions that recalled his days at the Naval Academy, Floyd once again became the student, intensively gleaning advice that the best astronomers in the world could offer. His first trip was to the Naval Observatory where he met Newcomb and observed through the 26-inch refractor. Newcomb, already disappointed about the delays in the project brought by Lick's change of his trust, was further worried by Floyd's lack of professional training in astronomy. Initially, he privately doubted that Floyd could carry out Lick's project, but nevertheless answered his questions and provided him with letters of introduction to astronomers in Europe. Floyd then headed overseas, accompanied by Cora and their four-year-old daughter, Harry.

Floyd found himself immediately embroiled in the reflector-refractor debate. In Ireland he met Howard Grubb, who for the Vienna Observatory had recently completed a 27-inch refractor which had nudged out the Naval Observatory's refractor as the world's largest. Grubb was admired throughout the United Kingdom for his ingenuity and technical expertise. Floyd then visited the estate of Lord Rosse, who had on his grounds a huge reflector made from polished metal. On a visit to Edinburgh, Scotland, Floyd talked with David Gill, who had earlier delivered his report on the two types of telescopes to the Lick trustees. Gill's final word to Floyd on the subject was to caution that it had always "been left to men of special mechanical genius to make and employ large reflectors."

Traveling to London, Floyd met with famed British astrophysicist J. Norman Lockyer, whom Joseph Henry had nominated to Lick as his choice for director. While there, Floyd also encountered Newcomb's assistant Edward S. Holden, who had helped prepare the first, rough plans for Lick's observatory. Holden was studying scientific instruments in England for the Naval Observatory and had timed his trip to London to coincide with Floyd's, both to ingratiate himself with the new trust president and to check on Floyd to see whether Newcomb's misgivings about him were valid.

Moving on to France, Floyd met with Urbain Leverrier, famous for his predictions of the existence and approximate location of Neptune, which led to its discovery by German astronomer J. G. Galle. Leverrier argued against building a reflector when he spoke with Floyd, claiming that no existing one worked well. Yet in Leverrier's own country, Floyd viewed a reflector built around a silvered-glass mirror in a new style that would soon prove superior to polished metal disks. (Added experience with this design came a few years later when George Calver successfully made a 27-inch silvered-glass reflector for English astronomer Andrew A. Common.) Floyd next paid a visit to the glassmaker Charles Feil in Paris, whom Newcomb had praised in his report to the Lick trustees. Floyd liked the price he quoted—$2,000—for the raw disks for a one-meter (40-inch) objective lens. He may have given this optimistic figure to help convince Floyd to choose a refractor.

While traveling through Europe with his family, Floyd kept up a constant correspondence with his contacts in America. He wrote the most to Henry Mathews, secretary for the Lick Trust. One of Floyd's first acts as president of the trust had been to hire Mathews as secretary; he kept the post for nineteen years until all the terms of the trust deed were completed. It fell to him, in one of his first acts as secretary, to wire Floyd the news of Lick's death, which occurred soon after Floyd had left America. During Floyd's European trip, Mathews struggled with the many legal difficulties besetting the trust, compounded enormously by Lick's death. He received instructions from Floyd, but there was so much delay between mailing questions overseas and receiving answers that he had to make many decisions on his own. Mathews was in charge, operationally, and he proved equal to the task. He had to oversee Lick's remaining properties, collecting rents and paying taxes to make sure that the estate did not lose its assets through negligence or oversight. Mathews was especially eager to keep the trust in good shape because of the legal battles brewing over Lick's estate.

Mathews' competence eased matters for Floyd, who had other problems to consider. Based on what he had learned from Newcomb and from his European tour, Floyd instructed the trustees to solicit bids for the manufacture of the objective lens from Grubb and from Alvan Clark. When he returned to America, he visited Clark in Massachusetts before coming back to California and joining the legal fray with Mathews and the other trustees. A serious contention had arisen that held up progress on the observatory for years. The terms of the trust stipulated that whatever money was left after Lick's many public works and his observatory were built would be shared equally between the California Academy of Sciences and the Society of California Pioneers. Naturally, the two groups were eager to receive their bequests quickly, and in as large amounts as possible. They were disappointed when, after Lick's death, the trustees upped John Lick's inheritance from $150,000 to $535,000 to quell his possible challenge of the estate. When the Academy and the Pioneers learned that this money was to come entirely out of "their" share, they decided to take the matter to court. Following a protracted series of legal maneuvers, the court decided in favor of the trustees. George Davidson, the long-time moving force at the Academy of Sciences, was undoubtedly disappointed by this episode on top of his earlier frustration with the development of Lick Observatory.

This dispute, together with other legal difficulties and many smaller claims against the estate that had to be fought, kept Floyd busy traveling between San Francisco, where he had business, and Kono Tayee, where his heart and wife wanted him to be. Floyd was to feel this pull for over a decade as he pursued the fulfillment of Lick's (and his own) dream. At the same time he kept up a constant correspondence on scientific matters. Floyd wrote that he felt he was spending all his time "pushing this infernal pen which I most heartily despise."

Finally, in spring of 1879 the trustees felt they had resolved the legal difficulties sufficiently to go forward with the project again. One of the first items they planned was a thorough investigation of Mount Hamilton's observing conditions, as Clark, Newcomb, and Holden had all advised. For this job they hired Sherburne W. Burnham, the noted double-star observer who was the advisors' unanimous choice. Then forty years old, Burnham was a completely self-taught amateur astronomer. Born and raised in Vermont, he had received a classical education through academy (the equivalent of high school), and then had taught himself shorthand to become a stenographer. He was a reporter for the Union army during the Civil War, an experience that surely must have provoked some interesting discussions between him and Floyd. While in occupied New Orleans Burnham had become interested in astronomy and had learned to observe with a small, inexpensive telescope. After the war he settled in Chicago, near Dearborn Observatory (which has since been relocated to the Northwestern University campus in Evanston, Illinois) with its 18.5-inch Clark refractor, the biggest in the world in the 1860s. (The telescope had originally been ordered by the University of Mississippi, but the Civil War wiped out the university's ability to pay, and the instrument ended up at Dearborn Observatory.) Seeing this telescope made Burnham want a better one for himself. In 1869 he met Alvan Clark, who was visiting Chicago, and ordered the best 6-inch refractor the Clarks could make. The telescope cost him $800, a large sum then even for a well-paid court reporter.

Burnham, short and wiry, had exceptionally keen eyes, which earned him his reputation among astronomers. He concentrated entirely on discovering and measuring double stars, pairs of stars that slowly orbit one another. By day, he worked in court to earn a living (he was married and ultimately had six children), and by night he kept busy with his telescope. With his Clark 6-inch, Burnham discovered over 400 new double stars, and published his results in scientific journals. At the time, astronomers were certain that the field of double-star observing had been played out, so Burnham's discoveries surprised and delighted them. His papers brought him to the attention of professional astronomers, which in turn led to his being allowed to observe as a volunteer with the Dearborn telescope. Using the powerful 18.5-inch refractor, he discovered still more binary stars, and began his long series of measurements of their orbits, from which their masses could be derived. By 1879 Burnham was one of the leading American experts in this specialized field of astronomy.

Double-star observing provided an easy way to judge Mount Hamilton's atmospheric conditions: what astronomers call "seeing." When the seeing is bad, the images of two close stars smear into a single tiny disk. The better the seeing, the easier it is to distinguish tight stellar pairs. Since Burnham was familiar with the appearance of many double stars under a variety of seeing conditions, he could use their appearance from Mount Hamilton to judge the wisdom of putting a telescope on the summit.

The Lick trustees wanted Burnham to spend a year observing at Mount Hamilton, but he could get away from his courtroom job for only two months. He agreed to bring his 6-inch telescope to California and observe with it there for a salary of $500, plus expenses. The trustees accepted this fee, and even authorized Burnham to order from Alvan Clark a new clock drive, the mechanism that tracks the telescope at the right rate to compensate for the apparent motion of the stars about the Earth's rotation axis. The trustees ordered a temporary wood and canvas dome for covering the telescope, based on Burnham's specifications. It was built in San Francisco in sections, then transported to Mount Hamilton and assembled by Fraser. Floyd wrote to ask Burnham what sort of personal shelter he wanted. Burnham replied, "As to other buildings, etc., anything will do. I am not afraid of roughing it. The obs[ervator]y and clock are the only important things."

By mid-July 1879, everything was in place at Mount Hamilton, but Burnham could not get away from Chicago for another month. He arrived in San Francisco via the transcontinental railroad on August 16.

His gear included his telescope, disassembled and packed in four boxes, another box full of the books he needed while observing (the Nautical Almanac undoubtedly among them), and "a few traps," or suitcases, totaling in all about 600 pounds. The day after his arrival he was established on the summit and noting the weather. "All clear & first-class seeing," was his first comment. He unpacked his telescope and had it temporarily mounted by that evening. Within a few days, Fraser had built a solid concrete pier for the instrument, and Burnham was observing on a regular basis. The first measurements he recorded were of the triple-star system α Scorpii on August 22.

Burnham had come prepared to like Mount Hamilton ("I am greatly interested in the place, and have but little doubt, unless there is something very peculiar in the general location, that the place will prove to be extremely desirable for astronomical work," he had written) but the reality exceeded his expectations. Night after night he recorded in his observing book remarks such as "First class seeing. No wind at all," and "Splendid night—absolutely still." He observed every night he could, except for one short visit to San Francisco.

In September, Newcomb came out from Washington and went with Floyd and Fraser to Mount Hamilton, where he observed with them and Burnham for one night before returning east. He was not as ecstatic about the site as Burnham, but his experience in critical observing was far less than the Chicago amateur's. In the end, Burnham found that forty-nine nights out of sixty were suitable for observing at Mount Hamilton, far more than there would have been in Washington, Chicago, or any other site then in use in America. Furthermore, Burnham called forty-two of the forty-nine suitable nights "first class," with exceptionally steady atmospheric conditions. During his stay he discovered forty-two new double stars, and picked up independently a great many previously known ones. As he wrote Floyd:

> If the proof of the pudding is in the eating, these discoveries ought to be eloquent on the subject of M[ount]t H[amilton] for the site of an Obs[ervator]y. They will certainly be much more satisfactory to outsiders than the mere opinion of anyone.

Burnham's site test did prove it was an excellent mountain for an observatory. Mount Hamilton offered not only a run of clear nights during his stay, but also a steady atmosphere that resulted in small, exceptionally bright star images. Burnham, partly because of his vacation schedule, but also no doubt as the result of quiet, careful planning by Floyd, visited Mount Hamilton during the best observing season of the year, the dry, clear months of late summer and early fall. Nevertheless, the summit was, and is, a good observatory site.

With the land already granted to the Lick Trust by Congress, and the road to the summit already built by Santa Clara County, it is hard to imagine what would have happened if it had been found a bad site for observing. Burnham's observing test gave everyone fresh enthusiasm, including Newcomb, who had earlier been discouraged about the project. After his two months on Mount Hamilton, Burnham was invited by the Floyds for a week's stay at Kono Tayee, a well-earned vacation for the sharp-eyed astronomer.

In April 1880, Floyd and Fraser traveled east to meet with Newcomb and Holden (the first Floyd had seen of Holden since London) and finalize plans for the observatory. Floyd authorized Newcomb to negotiate with the Clarks for construction of a 36-inch objective lens for a refracting telescope, and a contract was soon signed with the Clarks for the price of $50,000, considerably lower than Alvan Clark's first estimate to Lick. Newcomb also ordered for the trust a 12-inch refracting telescope ($6,300), a 4-inch comet seeker ($400), and a photoheliograph ($500).

Floyd and Fraser spent a month in the east with Newcomb and Holden refining the plans for the observatory until they were ready for delivery to a draftsman. This work was aided considerably by a detailed contour map of the summit that County Surveyor Herrmann had prepared. After his return to California, Fraser and a crew of three men headed for the top of Mount Hamilton, and on July 20, 1880, began preparing the summit. Their first chore, after hauling building materials to the top, was to raise a boarding house for the main work crew to come. Soon, a blacksmith shop, carriage house, hayshed, stable, and two small cottages were also erected near the boarding house. Livestock and chickens helped populate the summit. Fraser's wife, Floretta, joined him on the mountain and shared in the hardships of establishing a life on the isolated mountain.

With a mini-city in place, Fraser and a crew of thirteen men began blasting the top from Mount Hamilton's steep peak to create a level spot for the observatory's Main Building. "Every square foot of available surface [created] involved the removal of a prism of hard rock, one foot on the base and from 10 to 32 feet high," Floyd later wrote. Nearly a ton of black powder went to the job, and all told nearly 70,000 tons of rock were blasted free and moved by hand to clear the top. The work continued until December, when fierce winter storms blew in and forced a retreat of the work crew. Fraser and his men did not return to finish the job until the following April.

Fraser made two finds on the mountain that simplified work considerably. Exploring the mountain one day, he discovered a spring a short distance from the summit. Previously, the lack of a water supply on Mount Hamilton had concerned both Fraser and Floyd because water was essential for satisfactory living conditions and for the hard construction ahead. Hauling water up to the summit from Smith's Creek, seven miles down the road and 2,000 feet lower in elevation, was one of the daily tasks for the first workers. The supply Fraser found eliminated this drudgery. It provided 5,000 gallons of water a day in the spring and 800 a day in the dry summer months, enough for a thirsty work crew and for water power before electricity was available on the summit.

Fraser's other discovery, a bed of clay suitable for brickmaking, provided a source of good building materials two miles from the summit. A kiln was quickly set up. After test firings showed the bricks to be of good quality, the workmen began burning 10,000 bricks a day in the kiln. This on-site brick production greatly reduced the amount of material that had to be hauled up the mountain, and saved the trustees an estimated $46,000 in material and transportation costs.

With bricks available, Fraser and his crew raised a small building for the 4-inch transit telescope, and began laying the foundation for the dome to house a 12-inch refractor. Fraser began working on the dome in August, when the mountaintop was only slightly cooler than the scorched valley below, and finished it three months and some 32,000 bricks later. The rows of brick circled upward to form a cylindrical tower twenty-five feet in diameter and two stories high.

Alvan Clark thought the objective lens of the 12-inch refractor was one of the best he had ever made. The Clarks had originally sold the instrument to Henry Draper, a wealthy amateur astronomer in New York, but in 1880 he had exchanged it for an 11-inch photographically corrected refractor the Clarks had made. With this new telescope, Draper obtained in 1880 the first successful photograph of a nebula, a two-hour exposure of the center of the Orion nebula. Draper was happy, but so was Floyd because he got a quality 12-inch telescope from the Clarks at a secondhand price.

Work on the 12-inch refractor brought Floyd and Fraser in contact with Worcester Warner and Ambrose Swasey, partners in a midwest machinery firm. The two mechanical engineers wrote to ask about getting the job of manufacturing the metal track and rollers on which the dome supported by the two-story brick cylinder would turn. Floyd had intended to have the work done in San Francisco, but could not find a firm that could handle the job at the right price. Both Holden and

Burnham advised Floyd that Warner and Swasey did first-rate work and that he should hire them for the job. Floyd sent them the drawings he and Fraser had put together detailing the dome, and awaited their bid.

Fraser made the dome itself on the mountain, framing the hemisphere with narrow wooden beams steam bent and anchored into a curved shape, and covering this skeleton with a thin layer of plated-copper sheeting. Rolling up a sheet of corrugated steel uncovered a three-foot-wide slit through which the telescope would view the sky. Floyd deemed Warner and Swasey's bid acceptable and the trustees contracted with them to provide the metal track to top the circular brick wall, and the rollers, or "trucks," and gear to rotate the dome around the track.

The 12-inch telescope arrived in October 1881, and Fraser and Floyd mounted it on its pier, a brick column that stood in the center of the dome building. The mechanical requirements for the telescope were crucial. The instrument had to track the stars without the slightest tremor, which would be greatly magnified because the telescope focuses on such a tiny spot of sky and enlarges it greatly. Even small vibrations would make the object being observed bounce around wildly in the field of view. The telescope was mounted on a pier that stood completely isolated from the building walls so that vibrations from the turning dome would not be transferred to the telescope.

The dome itself was turned by a loop of steel cable that wrapped around the exterior and a wheel mounted in a recess in an interior wall; cranking on this wheel spun the dome around. Although fabricated by Warner and Swasey, the design for this equipment came from Floyd and Fraser, who had borrowed the idea from Howard Grubb. It was a new method for revolving a dome, and worked so well that it was later adapted at many other observatories.

At the beginning of November, Warner came to Mount Hamilton to supervise installation of the dome's rolling gear. No doubt Warner also hoped to meet Floyd and grease the wheels for getting the contract for the larger dome and the telescope to come, which Holden had told the two engineers about. Warner and Swasey formed an ideal team. Warner was the consummate salesman and knew what to expect from machine tools. Swasey also was expert at machining, but his greatest asset was figuring out how to turn ideas and design needs into finished products. The two New Englanders had apprenticed in the same shop and gone on to establish their own company, which operated in Chicago for many years.

Warner had first met Holden on a trip to the Naval Observatory to see the 26-inch refractor. The observatory was closed, but Warner talked Holden into giving him a tour, and impressed the young astronomer with his interest in manufacturing telescopes and other astronomical instruments, and his obvious familiarity with machinery. Holden later visited their factory, and arranged the sale of Warner and Swasey's first telescope, a refractor with a 9.5-inch Clark lens. Holden paid close attention to the engineers' progress with telescope designs, and supported them at every chance he could.

For their part, Warner and Swasey relied on the manufacture and sale of precision drill presses, mills, and lathes to keep their company solvent, but they looked at their telescope work as a way to show off their capabilities and bring acclaim to their company. In addition they had a genuine interest in astronomy. Warner was determined that their work on the 12-inch dome, along with his skills as a salesman, would put them in the running for the future work on the Lick telescope. Warner and Swasey kept up numerous rounds of correspondence with Floyd and Fraser on many nuances of the design and construction of the dome as a way of cementing the ties between them.

The dome installation went smoothly, and the telescope was operational in time for a transit of Mercury across the face of the sun on November 7, 1881. Holden and Burnham came to Mount Hamilton to observe this event. Burnham, as the most experienced observer, used the 12-inch refractor, Holden worked with a 4-inch telescope (the comet seeker Newcomb had ordered earlier for the trust from the Clarks), and Floyd witnessed the event with a portable, 2.5-inch surveyor's instrument.

Having multiple observers helped guarantee all would not be lost if a problem developed at one observing station, and also added reliability to the observations of this important transit event. Mercury and Venus, the two planets that orbit the sun interior to the earth, can occasionally be seen to cross the sun's face. Following an inside track, they each orbit in less than a year, and are always overtaking the Earth in the race around the sun. But because the orbits of the planets lie in different planes from that of the Earth's orbit, we do not see Mercury or Venus directly in line with the sun at every conjunction. Astronomers in the nineteenth century were eager to observe these infrequent transits. By precisely timing the appearance of the black dot of the planet on the sun's bright face from different points on the earth, they could calculate a figure for the Earth's distance from the sun, thus setting the length scale for the solar system. Other methods have since replaced transit observations, but at the time this was important work for providing fundamental information about distances in the solar system.

Holden reduced the Mount Hamilton trio's observations and reported them in a short scientific paper. Aside from the work Burnham had done while investigating the mountain's observing conditions, this was the first scientific result to come from the budding observatory. The transit also gave Holden and Burnham the chance to inspect the installation of the 12-inch refractor, which had been carried out by two non-astronomers, and the working of its innovative dome.

Although Mercury's transit provided an occasion for the first scientific use of the refractor, it was not the first time the telescope had been used to study the heavens from Mount Hamilton. That privilege fell to the King of Hawaii, when that island chain was still a monarchy. Years before, when Floyd had captained a steamer that traveled from San Francisco to Hawaii, he met King Kalakaua. When the king asked to visit Mount Hamilton, Floyd welcomed him warmly, even though it meant Fraser had to mount the 12-inch refractor temporarily in the unfinished building, which still lacked its capping dome. Thus, the first visiting astronomer to Lick Observatory was an enthusiastic amateur from overseas. He did not impress Fraser much, however. His log entry noting the king's arrival read "Capt. Floyd left to bring the King up today one load sand." Fraser, never a stickler for punctuation or proper spelling, seemed to equate the king's value with a shipment of building material. The next day, after the royal visitor departed, Fraser concluded "The King was almost the first to look through the 12 Inch last night So much for a King." For his part, the king was reported to have enjoyed his visit.

The second professional use of the 12-inch came on December 6, 1882, when a transit of Venus across the sun was observed at Mount Hamilton. Venusian transits occur as a pair eight years apart, but then not again for approximately 130 years. The first member of this nineteenth-century pair had been in 1874, and it was not visible from the United States. Expeditions to other parts of the globe had returned with unsatisfactory results, so astronomers were determined not to lose this next chance, the last until 2004. Floyd had hoped to convince Newcomb to come observe at Mount Hamilton for the prestige his presence would bring, but Newcomb decided instead to head an expedition to the Cape of Good Hope. Newcomb referred Floyd to David Todd, formerly his assistant at the Naval Observatory and now the professor of astronomy at Amherst College in Massachusetts. Holden, fearing any visit to Mount Hamilton by a potential rival for the directorship, advised Floyd that he and Fraser could do the work alone, "without any other celebrated astronomers. . . . The secret of success is to have the mountain to yourself so as not to be bothered." But Todd, perhaps just as anxious as Holden to push himself forward, persisted, and Floyd invited him to come. Todd might better have stayed at home. During his two months in California, his wife, Mabel Loomis Todd, back in Amherst, began "the rapture of her rare, intense [and adulterous] new friendship with

Austin" Dickinson, brother of the poet Emily Dickinson, which is chronicled in the book *Austin and Mabel.*

The Lick trustees paid for Todd's trip to insure that Lick Observatory, incomplete as it was, would be included in the scientific efforts. Todd traveled west accompanied by a photographer (photography was just starting to be effectively used in astronomy) to record the event. For this work, Todd used the photoheliograph, a five-inch telescope and camera for photographing the sun, which Newcomb had ordered. Made by the Clarks, the instrument had arrived shortly before Todd did. Floyd and Fraser assembled it and found that the instrument's focal length— set by the lens shape—was not what they had specified. They had to take a long series of test photographs to locate the instrument's focus. This was demanding work for two newcomers to astronomy, but they were successful and proud of their work. Thus, when Todd arrived he was soon set up to take a sequence of short-exposure photographs lasting throughout the transit. Floyd watched the event with the 12-inch, while Cora Floyd and Floretta Fraser marked the exact time of each observation.

After the photoheliograph, the next instrument to arrive on Mount Hamilton was the meridian circle, which was installed in 1884. The optics for this instrument came from the Clarks, but the precision mechanical mounting had been made in Germany. Fraser erected the brick piers to hold the instrument, and devised a roof he thought ingenious that opened upward the way a pair of swinging garage doors opens outward. Holden advised against Fraser's parting roof, but was later convinced by its easy operation and weather-tight seal that it was a good design.

Holden at this point was offering frequent advice via a constant barrage of letters to Floyd. He had suggestions for practically every aspect of the ongoing construction and purchases, often to the annoyance of Fraser. As the one on the mountain doing the actual work, Fraser resented some of the suggested directions from Holden, who was far from the scene and the realities of the job. Floyd, however, continued on good terms with Holden. He obviously welcomed having someone besides Fraser with whom he could discuss the observatory's overall direction, someone to help shoulder the responsibilities of the project. Floyd and Holden generated a friendship that lasted until the completion of the observatory drew near.

A firmer bond linked Fraser and Floyd, created by the demands of building the observatory as they and their men continued working hard on Mount Hamilton. After the dome housing the 12-inch refractor had been completed at the northern end of the leveled summit, the Main

Building began spreading south, brick by brick. Timber was hauled up the mountain for roof beams. Fraser ordered marble slabs from Vermont (he once wrote that Lick Observatory would have nothing but the best materials while he was in charge) for flooring the high-ceilinged hallways. Smaller, auxiliary buildings sprouted around the Main Building. From the valley floor below, the painted walls of the Main Building and the 12-inch's dome could be seen shining brightly in the California sunshine. But by 1884, Fraser and Floyd had done all they could at Mount Hamilton. They could not start the next stage of the project, which was to build the large dome that would house James Lick's great telescope. The dome's dimensions would be fixed by the telescope's focal length, but that would not be known until the Clarks received the glass disks to shape into the objective lens. The disks had been ordered in 1880 from Feil. In Paris, things were not going well—the House of Feil had been working on the order for four years without success.

4

The Telescope Comes to the Mountain
1884–1888

Far north of Mount Hamilton, on the shores of Clear Lake, Cora and Richard S. Floyd lived a life of leisure at their estate, Kono Tayee. Using Cora's inheritance, they built Kono Tayee to be not just a home, but also a memory of their childhoods. Its size and grounds recalled the plantations of their youths, and the large staff of Chinese servants allowed them the chore-free lifestyle that slaves had provided before the Civil War.

The spacious two-story dwelling stood out from the rustic backdrop of woods surrounding Clear Lake. A broad veranda, adorned with small cannons from Floyd's collection of weaponry, circled the house. A large central hallway, paved with marble and open to a skylight thirty feet overhead, led to high-ceilinged rooms—the library, dining room, billiard room, and parlor—each different and each elegantly ornamented. Visitors marvelled at the furnishings. With wall space enough to make a gallery owner jealous, Floyd was able to display many of the paintings and drawings he had created over the years.

Outside, tied to piers or anchored along the shoreline, a small navy bobbed in the Clear Lake waters. Floyd had built a steamer, the *City of Lakeport*, that daily went around the lake with mail and passengers. Next to this eighty-foot vessel rode *Hallie*, a smaller yacht also powered by a steam engine. Hallie was named for Harry Augustus Lyons, the Floyds' flamboyant daughter and one of the great joys of their lives. Filling out the fleet were several sailboats and other small craft. And when that armada did not satisfy the Captain's thirst for naval action, he would bring out model ships and sail them in a large fountain outside his home.

While the waters beckoned from one side of Kono Tayee, the woods stretching out on the other side offered an abundance of game. Floyd thrived on the outdoor life. He and Cora both enjoyed reputations as good shots, and spent many days together hunting and camping in the woods, which were still home to bears in the 1880s. "[S]he hunted deer and duck with the excitement of a keen and successful sportsman," is how a newspaper described Cora's efforts. Even when her husband was away, she liked to go hunting on her own.

Kono Tayee presented Harry with a setting for growing up that most children today can only imagine. Like her parents, she too spent much time outdoors, playing the tomboy to the hilt. Even as an adult she preferred pants to dresses, enflaming the gossips of rural California in the late 1800s. An only child, she enjoyed much attention from the adults around her. Far from the company of playmates her age, she created her own entertainment by acting out the stories she had read, creating costumes from the many heirlooms that filled the closets and attic at Kono Tayee.

Floyd's world at idyllic Clear Lake offered him nearly everything he wanted. He could pretend to a return to his pampered childhood, had full access to outdoor activities, enjoyed the company of his wife and daughter, and could entertain visitors in high style. If he and Cora tired of their isolated locale, they could head for San Francisco and take in the opera or join the many high-society parties. Floyd could have settled in to a relaxed time of recreation and socialization, but instead he chose to accept the headaches and responsibilities of carrying out the terms of James Lick's trust deed. Floyd's earlier life had been marked by action and command, and he had no doubt grown restless in his early retirement. Becoming president of the Lick Trust returned him to a position of authority, and he assumed the job with great enthusiasm.

Likewise, Thomas Fraser, Floyd's close associate in building the observatory, gained new importance in his position of superintendent of construction. Undoubtedly he worked for the Lick Trust from a sense of loyalty to his former employer. But he also had the eyes of astronomers around the world focused on his work atop Mount Hamilton. Prominent scientists, such as Simon Newcomb or Edward Holden, would visit the mountain and find Fraser there, supervising the project. People in the Bay Area and scientists worldwide watched the observatory's progress with great interest. Fraser and Floyd had seized their place in the spotlight, but not without paying the price of hard work. Floyd, especially, had to attend to innumerable problems.

Floyd could handle much of his correspondence from Clear Lake, but many observatory matters required his presence in San Francisco or atop Mount Hamilton. All too frequently for Cora's or his own plea-

sure, Floyd would make the long trip from Kono Tayee to Mount Hamilton, crossing Clear Lake by boat to catch a local stage, then transferring to a large stagecoach that would take him to Vallejo or
Calistoga. He could take a ferry from Vallejo, or the train from Calistoga to San Francisco. This part of the trek would take all of one
day—Mount Hamilton lay yet another day beyond. After traveling by
train from San Francisco to San Jose, Floyd would begin the winding,
twenty-six mile ascent of Mount Hamilton, by stage if his timing was
good, or on foot if it was not.

When Floyd knew he would be at the mountain for a long time, or
during the summer when the weather was good, Cora and Harry would
come to Mount Hamilton with him. The family lived in part of the
Brick House, a large dwelling just below the Main Building. For Fraser
and his wife, Floretta, the Floyds' presence meant a pleasant change in
company from the construction crew they saw daily. Sometimes they
played poker, while at other times Harry kept the adults entertained
with her great energy and sense of imagination. At Mount Hamilton,
the Floyds often found themselves hosts by default, for many visitors
who came to witness the construction would mistime their departure
and be trapped by darkness. Though their larders were limited and trips
to the store difficult to arrange, the Floyds, and the Frasers in their
turn, usually managed to scrape up some food and find a place for the
uninvited guests to sleep, even if only in the hay in the stable.

Floyd and Fraser could stand together and gaze at the brick walls of
the Main Building stretching across the leveled mountaintop, and the
dome for the 12-inch telescope standing at its north end. But turning
south, they faced the incomplete end of the building, which stopped
abruptly at a broad area prepared for the large dome. As much as they
wanted to proceed, there was nothing they could do until Alvan Clark
told them the focal length of the objective lens, which would set the
dome's size. The Clarks needed to receive both blanks from the Feils
before they could estimate the final, clear lens diameter, and from it
the focal length. One disk had come quickly, but the Clarks and the
Lick trustees in turn had to wait years for the other member of the
pair.

Charles Feil and his ancestors were part of the tradition that had
gained France its international reputation for glassmaking in the eigh-
teenth and nineteenth centuries. Several generations of noted glassmakers were behind the factory, or "house," that Charles Feil operated. His
grandfather, Pierre Louis Guinand, was the founder of the business. One
of the Feils' specialties, and one of the skills handed down over the
generations, was making high-quality flint glass, more commonly
known as leaded crystal.

The most familiar type of glass is crown glass, formed by melting a mixture of sand, limestone, and soda ash. This glass has small refractive power, which means it does not greatly alter the path of a light beam passing through it. Flint glass, however, is highly refractive, which causes the color and sparkle when this glass is cut and makes for its use in fancy glasses and crystals. Flint glass was discovered in the seventeenth century when a glassmaker added crushed flint to the sand rather than soda and lime. The resulting glass suffered from numerous microfractures. In subsequent batches the glassmaker avoided the fractures by adding lead to the mix. Later work showed that only the lead was needed for good results, and although flint is no longer used its name remains.

Feil's grandfather was known for turning out flint glass of exceptional clarity and sparkle. His methods were handed down generation by generation, and the Feil family earned awards throughout Europe for their production of flint glass for many purposes. In Charles Feil's day, the firm made a variety of flint glasses. Perhaps what most interested Floyd on his visit to the factory in 1876 was that Feil had cast the blanks for the 27-inch refractor at the Vienna Observatory, which in 1875 succeeded the Naval Observatory's 26-inch telescope as the world's largest. (Feil also later cast the blanks for the 30-inch telescope at the Pulkovo Observatory, built by Howard Grubb in 1885.) That success, the low price Feil had quoted to Floyd for the glass blanks, and Newcomb's recommendation of the company all insured that it be considered for casting the lenses. When in 1880 the Clarks contracted with the Lick trust to make the objective lens, the contract specified that they order crown and flint glass disks, 36 inches in diameter, from the House of Feil.

Charles Feil in the meantime had retired, turning control of the factory over to his son, Edmond. Feil's son apparently had learned the family secrets well, for he produced the flint disk without problem or delay. But the disk of crown glass eluded his attempts time after time. Just when success appeared to be at hand, the disk would shatter or turn out too full of fractures to be usable. The failures continued, one after another, and the months of waiting stretched into years, leaving Floyd and the Clarks in states of deepening despair.

To produce a disk, Feil would mix and heat nearly a ton of raw materials in a huge clay pot. As the mix melted it released gases, forming bubbles that had to be patiently stirred out of the brew. When the melted glass was sufficiently bubble-free, the pot was transferred to an annealing oven where the temperature would be very slowly decreased over a period of weeks, keeping the entire body of now solid glass at essentially the same temperature while it cooled. Annealing was essential to avoid fractures or stress patterns, which would occur if the outside of the glass mass cooled faster than the inside. When the annealing was finished, the pot was removed from the oven and the clay broken away, along with a considerable amount of the outermost glass. The workers would scan the misshapen glass lump for a chunk of material that was free of defects and large enough to press into a disk. If they spotted such a piece they would break it free from the larger mass and place it in a press. By slowly heating the glass chunk to the point at which it softened, it could be squeezed into a disk. If all went well at each step, the job would be finished.

For Edmond, failure would sometimes face him as soon as the glass came from the annealing oven; at other times the glass would shatter as it was being pressed into a disk. Depending on when in the process failure occurred, sometimes weeks, sometimes months were lost. The number of failures topped ten, then fifteen, and was climbing toward twenty when Charles Feil finally returned from retirement. The Clarks had received no satisfactory reply from their letters to Edmond asking for news of the disk's progress, so they turned to Charles for information. His response was discouraging. In October 1884 he wrote that he had attempted to take back control of the company, but his son would not let him. His son was drinking heavily and heading the company into ruin with large debts. "[Edmond] has allowed himself to be led by bad counsel, and today I consider his position as lost. His creditors will without doubt offer the house for sale . . . I shall be able perhaps to rebuy it . . . If I succeed in getting possession of my house, it will only take a little while to complete these orders," he said in the letter.

In February 1885, Charles Feil regained control of the factory. In May he wrote the Clarks that he had cast the crown glass. He cut away the defective parts, releasing a lump of stress- and fracture-free glass that he then successfully pressed into a disk. He finished the work and shipped the disk to the Clarks in September, ending five years of delay. Word of this success came by telegram to California, where the Lick trustees received it with great joy.

During the long wait for the glass, Floyd had experienced other frustrations. Many people in the Bay Area had grown just as tired as he had waiting for the observatory's completion. A group within the Society of California Pioneers—the same malcontents who had raised a fuss over John Lick's inheritance—anxious for their share of Lick's money and afraid that the trust's assets were dwindling, alleged that the trustees were mismanaging the estate. The Pioneers claimed that the delays in construction had wasted hundreds of thousands of dollars. Reporters used the opportunity to generate headlines. The attacks were frequent and personal, laying blame on all the trustees, but particularly on Floyd. The strident voices were finally stilled when the trustees showed that Lick's estate had actually increased in value over time, so that far from wasting his money they had protected it well. In the meantime, they were pleased to add, the public would have to learn patience, for building an observatory was not an undertaking that could be accomplished overnight.

At this point Floyd was feeling the full pressures of his position. He expressed his feelings in a letter to David Todd, professor of astronomy at Amherst: "I am really run to death with correspondence and business matters and I do most heartily wish that I was out of this Lick Trust business. I am worn out and disgusted with it and besides, it is a great expense to me, for which I gather only abuse and trouble."

Floyd acquired some moral support when Edward S. Holden came to the west coast in the summer of 1885. In 1881, Holden had left the Naval Observatory to become director of the Washburn Observatory in Madison, Wisconsin. When he did so, he had made it clear to Floyd that this was only preparation for his future role as director of Lick Observatory. Holden and Floyd had kept in close communication during the intervening years. Holden had continually urged Floyd to begin work on the dome for the large telescope, but the Captain resisted, saying that it did not make sense to begin construction until the telescope's focal length was known. The two remained on friendly terms, however, even though sometimes they were of different minds. Their similar military backgrounds gave them a broad common experience from which to communicate.

Holden moved west when he was offered the presidency of the University of California, a position he accepted with the understanding that he would assume the Lick Observatory directorship when the facility was finished. Holden had proposed that he take both posts at once, but Floyd adamantly maintained there would be no director until he had finished his work and turned the observatory over to the University of California Regents. When Holden took the presidency it appeared he would have only a few years to wait for the post he truly wanted. His optimism was buoyed when he learned that the glass disks had reached the Clarks' shop.

The Clarks were characters unequaled in the world of lensmaking. By this time, Alvan Clark was nearing the age of eighty, and his sons were only two decades younger. The three were much alike, thinly built and presenting dour expressions to the world. James Lick probably would have gotten along well with them. To the uninitiated, their shop must have appeared filled with mysterious tools and devices, and the

Clarks' appearance provided the setting's final touch for the arcane art of shaping raw glass disks into precision telescope lenses.

When the crown-glass blank arrived from Feil, the Clarks put it on a machine that spun it around in contact with a gritty surface. The rough, preliminary shaping quickly removed 300 pounds of glass from the disk. Then followed the steady, tedious work of rubbing abrasive on the glass by hand to slowly wear away the excess material and free the focusing shape within. By fall 1886, a year after the crown disk had arrived in their shops, the Clarks declared their work complete.

Floyd asked Newcomb to go to Massachusetts to test the 36-inch objective lens. Newcomb in turn asked Sherburne W. Burnham to accompany him, but he begged off, saying he was too busy, "[e]ven for the purpose of seeing the largest telescope in the world." Instead, Newcomb brought along Asaph Hall, who had replaced him at the Naval Observatory as the astronomer in charge of the 26-inch refractor. The Clarks had provided only a crude mounting for the lens at their shop, so it was difficult to aim and to test. The seeing, or atmospheric quality, was very poor during the entire week the two astronomers were there. Nonetheless, from the little observing they could do, they were impressed with the power and sharpness of the lens. It appeared perfectly acceptable, and they wired Floyd that it was ready to go.

Even though wintertime travel was risky, the Lick trustees decided to send Fraser east immediately to bring back the objective lens. Fraser arrived at the Clarks' shop and marveled at the two three-foot-diameter elements, the largest in the world. The large masses of glass carried a slightly smoky hue, and within their bodies floated a few tiny bubbles of entrapped air, minor flaws in otherwise perfect gems. Although broad, the disks looked delicately thin—each was barely two inches thick at its thickest. The front lens was shaped to be thicker at its center; the rear lens thinned at its center, and its back surface was flat. When held in a specially built metal cell the lenses would act together to bring distant objects into near-perfect focus. Seeing them gave Fraser a great sense of accomplishment, making him feel certain that the observatory on which he had been working so hard would become a reality.

For the trip back west, Fraser and the Clarks packed the two elements of the lens with meticulous care, wrapping each in many layers of cotton cloth, a thick layer of cotton batting, and finally paper for protection. The wrapped elements were packed in separate wooden boxes, which were in turn packed, surrounded and cushioned by curled hair, in steel boxes. These steel boxes hung on springs within still larger steel boxes, which themselves hung from wooden frames. The connection to the frames allowed Fraser to rotate the steel boxes. Every day on the bumpy train ride west he would tilt the boxes to a new angle, a precautionary move he and Floyd thought would prevent the glass molecules from being jiggled into an alignment that would have caused the objective to become polarized or otherwise transformed.

Fraser and the lens traveled on a private railway car, and their train was given special priority by the railroads, forcing other trains onto sidetracks until they passed. To avoid the worst of the winter weather Fraser and the precious glass disks went by a southern route, and made it safely to the station in San Jose. From there, he personally hauled the lens elements by wagon to Mount Hamilton. Fraser and Floyd installed the lenses in the metal cells that would attach to the telescope. They were assisted in this work by astronomer James E. Keeler.

Keeler, a promising young astrophysicist, had been hired by the Lick Trust specifically to provide scientific expertise as the observatory neared completion. He had been a student at Johns Hopkins University in its early days, and after his graduation in 1881 had gone to work as the scientific assistant of Samuel P. Langley, the director of Allegheny Observatory, near Pittsburgh. With one year off for graduate training in Germany, Keeler continued as Langley's assistant until, in the spring of 1886, Holden hired him on behalf of the Lick Trust to go to Mount Hamilton and work with Floyd.

Floyd agreed with Holden that the time had come for an astronomer's presence, but Fraser had his doubts initially, reflecting his proprietary feeling for the project. He wrote in his log, "I am afraid Capt R S Floyd has done the wrong thing in alowing a man to interfere with the construction, as an astronomer will. . . . it will however all be in the man if he has the right ring all will be right but if stubern then things will go wrong and he will have to leave that is all there is about it." Keeler's abilities and good-natured sensibility won over Fraser in the end, and though Fraser thought him "slow" he grudgingly admitted that Keeler knew what he was doing.

Keeler's first task at Lick Observatory was to set up a time service, modeled on the one that Langley had founded at Allegheny. Using the meridian circle, Keeler made precise star sightings that provided highly accurate measurements for setting the observatory clocks. The clocks, in turn, were used to send time signals over the Western Union telegraph lines along the entire Pacific coast, and east to Nevada. Every day, precisely at noon, a signal went out from Mount Hamilton to all the subscribers, who duly adjusted their clocks. The service was a valuable source of revenue to the observatory, and also helped to keep train travelers on schedule. Once the time service was operating on a routine basis, it required little of Keeler's time. He was then put to work preparing the manuscript of the first volume of the *Lick Observatory Pub-*

lications, a description of the instruments and a compilation of reports and documents back to the beginning of the Lick Trust.

When the 36-inch lens arrived on Mount Hamilton, Floyd and Keeler tried to measure its focal length in the long hall of the Main Building, but were unable to do so because they had no other large lens to use in the test. They also attempted to focus an image of the distant Sierra, over a hundred miles away from Mount Hamilton across the Central Valley of California, but were unable to shield the lens from the bright sky well enough to see the image. They then stored the lens in a safe in the director's office of the Main Building until the dome and telescope were completed.

Progress on the large dome had advanced far by this point. In 1885, the materials for the dome building that could not be made on site had been hauled to the summit. As soon as the Clarks had received the second disk, they could predict that the final focal length would be in the neighborhood of 55 feet, an accurate enough figure for work on the building to begin. One evening in early 1886, the Floyds, the Frasers, and the work crew gathered for a modest ceremony in which Harry Floyd laid the cornerstone for the dome. In the following days Fraser and his crew began laying out a circle 235 feet around, and then raising skyward, row by row, a thick wall of bricks for supporting the dome. Holden had retained Storm Bull, a professor of engineering at the University of Wisconsin, to provide a design for the building. His design included massive buttreses to anchor the round wall of the dome against hurricane-force winds. Floyd, on Fraser's advice, rejected this design, and the superintendent built the dome largely on his own.

Floyd came to the mountain with his family at this time, determined to be close to the scene of action during the final push to completion. By the summer of 1886, Fraser and his workmen had completed the outer brick walls. They next laid the foundation for the telescope's support pier. This foundation was constructed to serve a second, special purpose as well—to hold the body of James Lick.

On January 8, 1887, Fraser helped transport Lick's coffin from the Masonic cemetery to Mount Hamilton for a second funeral; it was not so heavily attended as the first. Lick had not specified how to dispose of his body, but when Floyd and George Davidson had separately suggested that he be buried beneath his large telescope he had offered no objections. The one thing he did object to was cremation, stating forcefully to Davidson, "I intend to rot like a gentleman." In simple proceedings atop Mount Hamilton, Lick was entombed in the foundation for the pier of the 36-inch telescope, thus finding his final rest beneath the monument he had envisioned years before. In deference to Lick's free-thinking ideas, no minister presided at the ceremony. Floyd delivered the eulogy, extolling Lick's virtues and munificence. Besides Fraser and the trustees, the witnesses included Davidson, Holden, Keeler, the mayor of San Jose, and members of the Pioneers. The coffin was opened so that those present could certify it was indeed James Lick's body being entombed. Davidson reported that "it was evident [Lick's] wish had been fulfilled." The coffin was resealed and lowered into a vault within the telescope's foundation, which was then capped with a huge stone slab.

If Lick had been able to return to life and look about him, he probably would have felt reassured that his trustees had been active, even though nearly a decade had passed since his death. Solid brick walls for supporting the dome rose around the foundation of the telescope. Above, the cylinder was open to the sky, but just weeks after the entombment, the iron girders to frame the dome arrived at the mountain, and by June the revolving roof was in place.

While the Clarks were grinding the lens, the trustees had issued calls for bids on the dome and telescope to selected companies. For fabricating and installing the dome, the trustees had chosen the Union Iron Works, a San Francisco company known for its manufacture of ocean-going steamers. From the beginning, Floyd had hoped to have the work done in the United States, and giving the order to a California company—when the state was still proving its mettle to the rest of the country—was even more appealing.

In discussing the dome for the 12-inch refractor, Newcomb had advocated a wooden frame, whereas Holden tried strongly to convince Floyd to build the dome of iron. The frame for the 26-inch refractor's dome in Washington was wooden, as were those of nearly every other dome existing. But larger telescopes require larger domes. The Naval Observatory's big dome was only 40 feet in diameter, while the hemispherical covering for the Lick telescope had to span 76 feet. With the Eiffel Tower approaching completion and offering full proof of iron's strength and versatility, it seemed to Floyd and the other trustees that building an iron dome was the best way to proceed.

Fraser had spent many months in the east in early 1885, touring all the major observatories and making notes on design concepts he thought worked and those that did not. Combining the best of what he had observed with ideas garnered earlier by Newcomb and Floyd, Fraser was able to produce nearly complete drawings for the dome. Fraser refined the plans with the manager at the Union Iron Works until they were ready to be handed over to a draftsman. The final design even accounted for the size changes that temperature variations induced in the metal track on which the dome would turn, an innovative touch for the times. The moving part of the dome weighed over 90 tons. The work cost $56,850, a price only about half of what other firms had bid.

For another $15,000, the trustees also contracted with the Union Iron Works for the framework for the elevating floor, a necessity given the range in height through which the eyepiece of the 55-foot-long telescope would travel. Pointed at the zenith, the telescope's focal point would be found near the base of the support pier; aimed near the horizon, its eyepiece would swing upward many feet. Rather than have astronomers perch precariously on a tall ladder to peer through the telescope, the 3,500-square-foot floor was designed to travel up and down seventeen feet to follow this motion. The wooden flooring covering the iron framework was worthy of Lick's own woodworking skills. Thin strips of light-colored cedar followed concentric paths, separated in parts by rings of dark woods. To elevate the floor, Fraser and Floyd had built a system of reservoir tanks on the high spots of the mountain, which they pumped full of water from the spring. Pipes carried the water downhill, generating enough water pressure to power the floor's rise (for as long as the water supply lasted).

Ireland's Howard Grubb had conceived the elevating floor. It was but one of his ideas that were used in the Lick telescope and observatory. Grubb's dream of making a large metal reflector for the trustees had been foiled when they decided on a refractor. Even so, he had hoped to make the objective lens, but saw that work go to the Clarks. Then, finally, he had lost out in the bidding for the dome and for the telescope itself. When the trustees were ready to contract for that work, they were under strong pressure to finish quickly. The time constraint and Grubb's distant location both worked against him in the bidding. All told, he lost out on several contracts worth many thousands of dollars, and instead received only a few hundred dollars from the Lick Trust as recompense for his ideas that were included in the dome and telescope designs.

This was a particularly bitter pill for Grubb to swallow, because he had believed from the start that the open bidding the trustees had specified on Newcomb's recommendation was demeaning to his own artistry and integrity as a telescope maker. The trustees reviewed the bids and selected the one they judged best, but reserved the right to cull ideas from other bids as they saw fit. The trustees had the best of all worlds— they could choose the manufacturer they wanted and at the same time incorporate worthy ideas from the losing bidders.

The winners in the bidding for the telescope were Worcester Warner and Ambrose Swasey of Cleveland, Ohio. At $42,000 their bid was the highest, but the trustees had the most faith in their ability to do the job well. Since undertaking the work on the dome for the 12-inch refractor they had gained more experience with telescope design and construction. The small instruments they had made were showcases of inspired ideas, some their own and others taken from their rivals. Although they were not always original, their expertise at machining enabled them to put into practice concepts that others could only envision. Warner and Swasey approached a telescope as but another type of machinery with its own special requirements. This attitude appealed to Fraser and Floyd's own mechanical aptitudes, and stood against the astronomers and artisans who felt telescopes were novel and unique creations.

Newcomb had once written to Floyd that "every increase in [telescope] size adds a new problem of how to make it work conveniently, smoothly, and successfully and follow a star without vibrating." Newcomb's admonitions were especially appropriate for the Lick telescope. Even though its aperture was only six inches greater in diameter than the largest existing refractor, that was a substantial step upward in size and weight for the telescope. The problem facing Warner and Swasey was to manufacture a cylinder of metal 55 feet long, 4 feet in diameter, and 6 tons in weight, and to make it move with the precision of a Swiss watch, offering the astronomer a smooth and convenient operation.

In the matter of convenience, Warner and Swasey had refined a system of "eye-end controls" for unlocking and locking the telescope in position with a twist of the wrist. The control knobs, distinctively shaped so they could be easily identified even in a dark dome, were located next to the telescope's eyepiece where an astronomer could quickly get to them. The controls replaced an earlier, cumbersome system of ropes that astronomers tugged to release and secure their telescopes, a system that was imprecise and hard to keep straight in the dark.

Grubb claimed to have invented eye-end controls, and he certainly had used them on the Pulkovo Observatory telescope. But Warner had first encountered the idea on a small telescope he had seen in 1876. Even though Warner and Swasey did not originate the idea, they had recognized its worth and put it to good use. They were able to do so, in part, because of Swasey's expertise at designing transmissions and gear-trains. Each control knob was at the end of a long rod, which transferred the knob's turning to a gear's rotation. The gearing effected the clamping or loosening of the telescope, and also controlled small motions to fine-tune the telescope's aim. A similar arrangement used the telescope's motion to turn dials located near the eye-end that indicated the telescope's position. All this made it possible for an astronomer to have complete control over the telescope from the observing position.

Burnham, visiting the Cleveland shop at the request of Holden, got to see these controls up close, and was impressed by the amount of effort that had gone into making telescope operation easy for the astronomer. He described his impressions to Holden: "Warner has so fully considered these things that there is not much room for suggestions— from a theoretical standpoint. It was evident that they had encountered many problems in matters of detail which had not been and could not be anticipated, and that some of these required a great deal of thought to work out a satisfactory solution."

Swasey's skill at gearing was even more important for making the telescope track smoothly. A clock-drive, similar to the mechanisms that ran old grandfather clocks, used hanging weights connected to gears to track the telescope across the sky at a fixed rate. As the weights, totaling 810 pounds, slowly descended through their travel of about 17 feet, they turned the gears at a steady rate. When the weights reached the bottom of their travel, which took a little over two hours, the astronomer would haul them up again. The system generated about three thousandths of a horsepower for moving the telescope, a small amount of power for such a large instrument and proof of the good job the engineers had done. Warner and Swasey had also made the weight system adjustable so that the clock-drive could be tuned to match closely each object's motion across the sky.

Powered by the clock drive, the telescope turned about its polar axle, aligned parallel to the Earth's rotation axis so that a simple rotation of the axle at the proper rate would compensate for the Earth's rotation and keep the telescope's view fixed. The axle would bear most of the telescope's weight, and Warner and Swasey had to make several castings of high-carbon steel before they succeeded with one they considered strong enough.

While the polar axle allowed the telescope to turn in an east-west direction, a pivoting cross-piece perpendicular to the axle, with the telescope on one side and counterweights on the other, created the telescope's north-south movement. These two motions could aim the telescope at any point above the horizon. Warner and Swasey spent much time at their foundry in making parts strong enough for this work also.

Another major problem they faced was to make the telescope tube sturdy enough so that its ends would not sag appreciably, which would throw off the optical alignment. Warner and Swasey made the tube in sections to be transported to Mount Hamilton and riveted together there. Following a suggestion from Holden, they tapered the ends of the telescope, reducing the tube's weight and consequently its sag. The telescope was 4 feet wide at the center, where it attached to the mount, but slimmed to 38 inches at the lens end and 36 inches at the eyepiece end. Floyd further resolved the sag problem when he suggested adding a set of counterweights that could be adjusted to keep the flexure at its maximum amount. If the telescope sagged the same amount at both ends at all times, its optical alignment would hold.

The trustees signed the contract with Warner and Swasey in the spring of 1886, with the hope that they would complete their work by early 1887. But as with many parts of the project there were unforeseen delays. Because of this, Fraser had left Mount Hamilton before the telescope components arrived. Seeing his part in the observatory drawing to a close, Fraser had committed himself to manage a property in southern California starting in November 1887. His departure marked the end of seven years on Mount Hamilton for him and his wife. The observatory owed much to his expertise and sense of dedication. As Floyd wrote, "Once Fraser gets his head fixed right, nothing will deviate him from the greatest care and exactness." Having been so close to the project for so long, Fraser undoubtedly felt mixed emotions when he left the mountain just before the telescope's arrival.

In October 1887, Floyd traveled to Cleveland, where he met with Newcomb and Burnham to inspect the work at Warner and Swasey's shop. The engineers had assembled the telescope and mounting to check their performance. The inspection impressed the trio greatly, not just for the telescope's performance but even more for its enormous size. Floyd likened the telescope to a huge mariposa pine.

Warner and Swasey continued making final adjustments until November, when they disassembled the tube and mount and shipped them to California. Swasey came with them to oversee their installation. From the train station in San Jose, teams of horses dragged the heavy wagon loads—some individual telescope pieces weighed several tons—up the mountain through the treacherous winter weather. From above, looking through binoculars, Floyd could follow their slow progress. Some loads tired the horses so greatly that they could not finish the trek in one day, and Floyd would worry through the dark hours until the journey could begin again. All the parts reached the mountaintop by December 12, 1887. Floyd and Swasey supervised the workmen who attached the mounting to the pier, and then secured the tube to the mounting. The work progressed quickly, but the weeks of hoisting the massive pieces into place were a time of great concern for Floyd.

I have been very busy since my return . . . at work from 6 a.m. until bed time every day without a moment to spare. . . . I could not help feeling anxious on account of the very disastrous consequences if any of these weights from 3 to 4½ tons, from an unseen flaw in a hook, or a careless lashing, had dropped 37 feet on to our elevating floor. It might have meant 50,000$ cost and a year's delay to the Observatory.

A witness to some of this work was Alvan G. Clark, the son of elder Alvan Clark. He was at Mount Hamilton putting the finishing touches on a photographic corrector lens to supplement the telescope's objective. Realizing belatedly the growing importance of photography in astronomy, the trustees had ordered this lens from the Clarks. The telescope's objective lens was intended to aid the human eye, and had been shaped so that it best imaged yellow light, to which the eye is most sensitive. Photographic emulsions, however, are most sensitive to blue light, and the photographic corrector was designed to fit in front of the 36-inch objective and make the telescope bring the blue light to a pinpoint focus. Clark's work on the mountain was slowed because of the bitterly cold weather; even with stoves going in the workroom the glass was too cold for the final polishing, and he lost much time waiting for weather warm enough to allow him to continue.

Meanwhile, the telescope quickly took shape within the dome. By December 31, it was ready for the objective lens, which was removed from the safe and carefully fitted on the telescope. Swasey, Floyd, Clark, and Keeler, along with members of the work crew, gathered in the dome for a first look through the telescope, hoping to greet the new year with the telescope's first light. But the howling wind and freezing rain and snow prevented them from opening the dome that night. As luck would have it, California was experiencing one of its worst winter storms in decades. Floyd described the blow as a "rattling South Easter," and estimated that the winds were moving at seventy-five miles an hour. The temperature stuck near zero for several days, and the firewood was so wet it was difficult to get a warm blaze going. "We are all rather uncomfortable and rather short of blankets," Floyd said in a message to town.

Waiting for the weather to clear gave Floyd time to explore a problem discovered while setting up the telescope. Warner and Swasey had made the support pier about a foot too high, he claimed, which shifted the objective of the telescope dangerously close to the inner surface of the dome. Floyd inspected the telescope and found this was not a fatal flaw; even with the addition of the photographic corrector, the end of the telescope cleared the dome by nine inches. Having built the dome slightly oversize accommodated the mistake.

The nights of January 1 and 2 passed under cover of storms. Finally on the evening of January 3, the sky cleared enough so that they could try to look through the telescope. The dome shutter was difficult to open, and the dome itself was impossible to turn—water had worked into the track and frozen the dome tight. The shutter was also frozen, but with effort the workmen cranked it open. Through the slit they could aim the telescope at whatever star passed within the shutter's view. They tried for bright Aldebaran, in the constellation Taurus. When they did, they found to their horror that the telescope would not focus. The star remained a large, circular smear in the eyepiece. They quickly realized the cause of the problem—the telescope was too long, and the eyepiece holder, or draw tube, could not be inserted far enough to draw the star into focus. Swasey himself sawed six inches off the tube, which solved this problem. The observers got Aldebaran in view and focused ("a blazing red sun," Floyd called it) before the clouds again closed over the mountaintop.

Snow fell for the next several days, suspending any chance of observing but giving the livid Floyd the chance to detail in writing the mistakes made by the Clarks and by Warner and Swasey. The pier height was solely the engineers' mistake, he claimed. And although Warner and Swasey had also built the telescope tube too long, the fault for that error lay with the Clarks, who had specified the focal length incorrectly. Floyd found both errors inexcusable, but fortunately they did not derail the telescope. It would have been worse, for instance, if the Clarks had given the focal length as too short, for the machinery was not on the mountain to make longer draw tubes. Of course if Newcomb and Hall had accurately measured the focal length when they tested the lens in Cambridgeport, or if Floyd and Keeler had done so when it was first delivered to Mount Hamilton, the fault could have been remedied long before first light, but Floyd did not mention these omissions in his hot memorandum.

The weather cleared again on January 7, giving them another chance to judge the telescope's power. Even though the dome still did not turn, the group could observe whatever objects drifted into view. In jittery handwriting, caused by the cold working on his ungloved fingers, Floyd wrote "We are all waiting in this office (next the big Dome) for Saturn to come by our shutter, which will be in about 2 hours." When Saturn arrived at about midnight, the group gave up the relative warmth of the office for the frigid dome interior. The sight of the ringed planet rewarded them for their patience. "The definition was exquisite," Floyd wrote, "[Saturn] had the silvery brightness of the moon. All hands were delighted. . . . There is no doubt that we have the most powerful optical instrument in the world."

The telescope's light-gathering power proved itself then by providing details never before seen in some of the observed objects. Keeler found a new narrow dark gap among Saturn's bright rings, a discovery due not only to the sharp images provided by the powerful telescope and the dark, clear sky above the elevated observatory site but also to Keeler's sharp eyesight. And when the group looked at the Orion nebula, Clark saw for the first time a faint, sixth member of the small group of stars known as the Trapezium at the nebula's center.

"The big telescope works well, I can safely say excellently," Floyd felt. "[B]ut there is a world of nice adjustments to be made before we get it in that shape that will do it justice and with such a monster every one of these means lots of climbing to great and giddy heights and great precautions to save one's neck."

Most of the adjustments were standard problems of alignment that any new telescope requires, but other deficiencies pointed to more serious problems. Swasey and Floyd isolated a source of vibration in the pier that caused the telescope to shudder while tracking, causing images to disrupt into a blur. The astronomers and engineers were uncertain how to solve this problem, though Floyd later advised that the hollow pier be filled with brick and sand to damp out the vibration. He also thought that because the metal support pier changed temperature through the day and night, its expansion and contraction would affect the telescope's tracking. Following Floyd's advice, the pier was later sheathed in wood to insulate it and reduce the temperature changes.

The other obvious problem with the telescope was that the degree of flexure considered acceptable for visual work with the telescope proved too great for photography. Ultimately these were minor problems, but they erupted in the local newspapers in a major way. Clark finished his work on the photographic corrector in February, and soon thereafter began critizing Floyd, Warner and Swasey, the telescope, the trustees, and everyone involved in the project except himself.

Clark, who according to Keeler was "a terrible old blow and a grumbler," had complained chronically while at the mountain, and his whining continued after he left. Unfortunately for Floyd, Charles Burckhalter was within earshot of Clark's complaints. Burckhalter operated the Chabot Observatory, in Oakland, which featured a 10.5-inch Clark refractor. He also had many newspaper contacts, and fed Clark's complaints to a reporter. The resulting story in the *San Francisco Chronicle* described the telescope as unusable and a complete failure, and laid blame directly on Floyd.

The Captain quickly guessed how the story arose, especially after he found out that Clark had met Burckhalter shortly after leaving Mount Hamilton. Though Floyd did not know Burckhalter, he supposed the Oakland astronomer held a grudge because he had not been included in the first mailing of the *Lick Observatory Publications,* and was exacting his revenge through the newspapers. Floyd let fly his feelings toward Burckhalter in a letter he wrote for inclusion in the observatory's records: "I regret that such is the specimen of the California Amateur Astronomer, more than I do the misfortune of having lived 44 years on this planet in utter ignorance of his existence or of his own '10½-inch' and so giving the offence that visits upon me his mighty vengeance."

Floyd also gave Clark a broadside, saying that he was a person who would criticize anyone's work that was not wholly his own, even his brother's. Clark's animus toward Lick Observatory and Floyd came in part because of his curmudgeonly personality, but it rested also in his jealousy of Warner and Swasey. The Clarks had hoped to build the entire Lick telescope, not just the lens, but they were not considered for this work. At one time, Warner and Swasey had approached the Clarks about entering a partnership to produce a commercial line of telescopes, with the opticians doing all the lens work and the engineers handling the mechanical construction. The Clarks turned down the offer. Warner and Swasey struck the deal instead with John Brashear, an optician in Pittsburgh who was the Clarks' chief competitor for most major astronomy products in the United States. Seeing his firm's position as the country's chief telescopemaker threatened, Alvan Clark struck out by criticizing the work of his rivals.

Burnham offered Floyd some words of solace when he heard of the newspaper reports. "You must bear with as much philosophy as you can muster the fool things which have been said about the Obs[ervator]y. . . . These things will soon be forgotten, while the monument you have erected on Mt. Hamilton will endure for all time; and will reflect more and more credit on your untiring energy, faithfulness, and devotion in carrying on and completing the construction. So much for the demnition bow-wows," he wrote. His last words sum up his feelings for the press; they had also been used by Charles Dickens to describe a character who was going to the damned dogs.

Floyd received much support from Burnham and others, and it helped him weather this blow. Still, the criticism hit him hard and came at a time when his health was failing. His heart was giving out, and the final push on the mountain to finish the observatory was proving too much for him. He was only forty-four years old, but the last ten years dedicating his life to completing James Lick's dream had taken their toll. Floyd stayed on the job until he considered the observatory ready for delivery to the University of California Regents. Then he followed his doctor's orders, and in early April retreated to Kono Tayee to rest and recuperate, never again to see Mount Hamilton.

Several days after Floyd had departed, representatives of the regents, accompanied by Holden, Swasey, and trustee Edwin B. Mastick, visited Mount Hamilton to inspect the observatory. Timothy Guy Phelps, who headed the regents' Lick Observatory Committee, doused the marble floors in the Main Building with tobacco juice. Keeler, who had by then been working at the observatory for nearly a year, gave the group a Cook's tour, and delighted them that night with views of Saturn and Mars through the 36-inch refractor. The next day they went down, satisfied with all they had seen. Holden came to the mountain on May 1, and began listing final items that the trustees should supply for the observatory. An inventory of all the observatory's property (from one 36-inch telescope down to one soup ladle) was drawn up and sent to Kono Tayee for Floyd's signature. The other trustees also signed it on its return.

On June 1, 1888, nearly twelve years after James Lick had died, members of the university regents and their associates met the Lick trustees, except for Floyd, in their San Francisco offices. The title papers had been drawn up, and were now signed and witnessed, legally transferring the observatory and all of its property to the University of California Regents. That property included a residual of $90,000 in gold from Lick's original bequest. The regents used this money to establish a fund to help operate the observatory.

The next day Fraser conveyed the papers to Holden in his now official office in the Main Building atop Mount Hamilton. Fraser hoped to offer a short speech in the dome of the Great Refractor, as the 36-inch telescope was called, to praise Floyd's efforts. Holden allowed Fraser only the private confines of his office for the brief speech. With that, the deed was done, and Holden had what he had long awaited. Now he, not Floyd, was the skipper, and Lick Observatory was about to set sail with a telescope "superior to and more powerful than any yet existing" and an excited staff of talented young astronomers.

5

Into the Ocean of Science
1888–1895

Trust president Richard S. Floyd, always one to turn a nautical phrase, had hoped "to launch [Lick] Observatory into the ocean of science," but the Captain was far from the scene when the trustees presented the observatory to the Regents of the University of California. With the transfer completed, Edward S. Holden took charge of the world's most powerful research station for astronomy. Its combination of telescope size and mountain site gave the observatory the lead over any facility then existing.

As its director, Holden had reached the pinnacle of his astronomical career, a remarkable achievement for the forty-two-year-old scientist. His name had been mentioned for the Lick directorship since 1874, when he had been a practicing astronomer for barely a year. His good fortune came about because he had befriended and impressed Simon Newcomb, whose recommendation of the young astronomer carried considerable weight. Over the years, however, Newcomb's views had apparently shifted. Just before Lick Observatory was completed, Newcomb supported David Todd, professor of astronomy at Amherst, over Holden as the best choice for director. Newcomb told Floyd he thought the post was too much for someone as young as Holden, even though ten years earlier he had considered him well qualified for the post.

Years later, following the difficulties that developed during Holden's administration, Newcomb apparently felt a need to defend his original selection. He did so, damning Holden with the faintest of praise, by writing, "I could not, at the moment, think of anyone decidedly preferable to him." Nor could the regents when it came time to choose. By then Holden was too deeply entrenched to be passed over, and they ignored Newcomb's suggestion of Todd for the directorship.

From 1874 until 1888, Holden had groomed himself for the Lick Observatory directorship. After Asaph Hall took over the Naval Observatory's 26-inch refractor from Newcomb, Holden had continued as Hall's assistant for a time, but then had transferred back to the meridian circle. His lighter observational duties there allowed him time to take charge of the observatory's library, work that let him exercise his true calling. Holden was a born cataloguer. In 1881 he was cataloguing books and observing star positions when he was offered the job of director of the Washburn Observatory and professor of astronomy at the University of Wisconsin, in Madison, which operated the observatory. This offer also came about because of Newcomb's recommendation. Going from the Naval Observatory to Washburn, with its 15-inch Clark refractor, was in some ways a step down scientifically for Holden, but in other important ways it was a large step upward. At Washburn, Holden was no longer following the orders of another astronomer. As observatory director he had full access to the telescope and other observatory resources, and he could plan his research accordingly. Other astronomers presented with similar opportunities have used them to make their research and scientific careers take off dramatically. Holden's continued to chug along in uninspired fashion.

Holden was very good at synthesizing information. He could adeptly assemble many disparate facts into a single cohesive body. In fact, this was his trademark as a scientist. While at the Naval Observatory he wrote a 221-page paper on the Orion nebula, a huge cloud of star-forming gas and dust. He ransacked the literature of the previous two centuries and compiled every observation and theory about the nebula that he could find. Tacked on briefly at the end were his own observations of the nebula made with the 26-inch refractor.

Holden attempted to make photometric, or brightness, measurements of regions in the Orion nebula by comparing their appearance with the image of an illuminated piece of paper simultaneously projected through the telescope. By adjusting the illuminating lamp until the paper matched the nebula, Holden could derive a numerical value of the brightness of the region in the nebula he was observing. Over a period of two years he thought he detected changes in brightness; these were never confirmed by other studies. Holden's study was probably flawed by the difficulty of accurately controlling the lamp. His inability to realize the problem points to some failure as a scientist, for whom full awareness of experimental errors is all-important for producing correct results. Holden's study, if verified, would have provided some insight into the then unknown nature of the Orion nebula.

The universe was still little explored and even less understood in the late nineteenth century. At the time of Lick Observatory's dedication, astronomy stood poised to make a great leap forward in understanding. Holden and his contemporaries knew that the sun and planets were part of a large system of stars, the Milky Way, but they knew of nothing beyond our galaxy's borders, or indeed of its nearby regions. In their telescopes, the wispy images of gas clouds in our galaxy were as mysterious as the faint blurs of galaxies external to our own. But new observing technology and laboratory understandings were becoming available at an accelerating pace, providing means for new insights into the heavens. Astronomers were just beginning to experiment with spectroscopy, the tool that would enable them to begin sorting out the myriad objects in the night sky.

Even while beginning to puzzle out the distant denizens of space, astronomers were still making major discoveries within the solar system, one of which took place while Holden was at the Naval Observatory—or rather, away from it. Hall had developed a plan to search for moons around Mars with the 26-inch refractor. The red planet was nearing opposition in 1877, placing it favorably in the sky for observing, and coming to one of its closest approaches to Earth in years. Hall thought he had a good chance of discovering whether the planet possessed any moons, but he faced one major problem—Holden. To reap sole credit for any discovery, Hall wanted to do the work while Holden was away.

The stars were shining favorably on Hall as Mars came into opposition, "and by the greatest good luck Dr. Henry Draper invited [Holden] to Dobb's Ferry at the very nick of time. He could not have gone much farther than Baltimore when I had the first satellite nearly in hand," Hall later wrote. News of the discovery spread quickly, and astronomers everywhere turned their telescopes to Mars and its moons. Holden and Draper did likewise, using a telescope at Draper's private observatory. They found what they for a time thought was yet a third moon, and over several nights performed every test they could imagine to rule out the possibility that the image was a faint background star, or a "ghost" formed by a stray light ray within the telescope. Holden wrote Alvan Clark about the discovery, giving the purported moon's distance from Mars and orbital period, and asked Clark to look for the moon and "settle it one way or another."

It turned out that this "discovery," in fact, involved sightings of a background star and on one occasion Hall's own newly spotted moon. Undaunted, Holden continued the search when he returned to the Naval Observatory, and again found what he thought was a new moon, and announced its position and period. His haste again got the better of him, for this time the moon he described had an impossible motion according to the equations governing orbiting bodies. Unfortunately, he reported the finding to his colleagues before he checked his figures and realized the "moon's" impossibility. Traveling by word of mouth, the oversight became common knowledge among astronomers and haunted Holden for years.

It did not hurt Holden's chance for advancement, however, and he received the job offer from the University of Wisconsin. When he moved to Madison and took up the directorship at the Washburn Observatory, he worked hard on a program of determining fundamental star positions, similar to his work with the meridian circle telescope in Washington. The highlights of his Wisconsin tenure, however, came in 1883, when he headed an expedition to a remote island in the Pacific to study a solar eclipse. The round-trip journey covered three months and over 12,000 miles.

Back at home, Holden kept active in the circle of faculty members at Madison, and cultivated his friendships with important men in the administration and among the regents of the university. Holden made a favorable first impression on most people. He stayed abreast of all the latest thinking and findings by astronomers, and in addition read widely from books outside of astronomy. The diversity of his interests is apparent in the subjects covered in the many books and articles he wrote: Mogul emperors, Aztec codegraphs, cryptology, bastion fortifications, and suicide among rattlesnakes, to name but a few. In conversation he could be entrancing, and his letters were always lively. His conversational skills and wide-ranging knowledge helped him favorably impress acquaintances in many situations. These talents also proved valuable in 1885 when he moved to California and accepted the presidency of the state's young university.

Given the qualifications then thought necessary for a university president, Holden's appointment seems strange, for he was not a moral philosopher nor had he held any comparable position up to that time. It was, in part, a measure of how small the university was, how prestigious Lick Observatory was to the university regents, and how strongly they and the Lick trustees considered Holden the person to head the observatory. The presidency was a plum they offered Holden to lure him west, and it assured his ascendancy to the head of the observatory.

The university's newspaper, *The Berkeleyan*, described Holden on his arrival as "above the average height and yet not excessively tall . . . [T]here was something in the wide-awake, yet quiet, that confident, yet not haughty, demeanor, something in the low, even voice, something in the wonderfully frank individuality that won the listener's respect. That something was indicative of self-control, of decision, of the ability to mark out a line of action and of the will-power to stick to it." Whether Holden was marked by as many contrasting qualities as they indicate is not clear, but the paper described him accurately in noting his willpower.

Holden used his opening address to stress that as university president he had a "double capacity," for he considered the Lick Observatory his most important charge. He also took the time to admonish the students in advance about their behavior. "The good name of the University is largely in your keeping. A thoughtless though insignificant action on your part may, when seen in the light of the public without, work a serious injury." For all his time as president he had trouble understanding the pranks that college students of his era pulled; the straitlaced president equated mischief with criminality, and he responded with strict discipline.

During his three years in Berkeley, Holden continued to demonstrate his personal charms in dealing with the university faculty and administrators. He was generally regarded as a competent president and was dutifully inspiring. He commented that the football team was a good rallying point for the university, and, even though expensive, it was worth it. Holden's position brought him into contact with many wealthy and influential people in the Bay Area. After he became observatory director he did his best to use these contacts as sources of gifts to support work or purchase equipment.

Holden's major astronomical effort during this time was to consider who among the astronomers active in America he wanted to make up the initial observatory staff. The group he chose consisted of S. W. Burnham, E. E. Barnard, James E. Keeler, and John M. Schaeberle. Holden made job offers to these men during the months when construction was nearing an end, and had their agreements to join the staff well before the observatory was completed.

For Burnham, joining the Lick staff meant returning to the Mount Hamilton site he had scouted years before, and the skies he had described so glowingly to the trustees. But he did not come easily. In many letters, Holden cajoled and persuaded the fifty-year-old astronomer to agree. Burnham was still employed as a court reporter in Chicago, enjoyed his job, and thought his astronomy would suffer greatly if he took it up as a full-time profession rather than a nighttime hobby. Also, he had to consider the needs of his wife and children, for whom Mount Hamilton could provide only inadequate housing and no schools.

Holden persisted, eager for the prestige Burnham's reputation would add to the observatory. Eventually he convinced the famous double-star observer that the prospect of using the great refractor from the splendid mountain site was an opportunity that far outweighed his personal reservations. And so Burnham came west to Lick Observatory. He housed his family in San Jose, and began working two nights a week on the

36-inch telescope, often using the 12-inch on other nights, a wealth of telescope time granted him on the basis of his senior position and his years of successful observing. He continued to discover, measure, and catalogue double stars, the object of his entire astronomical career. Although some double stars are merely chance alignments of two stars in the sky, the majority are systems in which two stars orbit one another under control of their mutual gravitational attraction. Determining the orbits of such systems allows astronomers to calculate the masses of the stars, providing the only direct means of measuring these important data. From these measurements come all estimates of star masses, including indirect inferences for solitary stars. Burnham's work thus formed an important underpinning for a very broad understanding of stars.

Burnham was also glad to lend his talents and the knowledge he had gained over his years of observing to help the younger members of the Lick staff. In particular, he and Barnard formed a very close friendship. Burnham probaby saw signs of himself in the self-taught young astronomer, who was driven, like him, by his love for working with a telescope. And Barnard looked to Burnham in part as a replacement for his father, who had died before the young astronomer was born. Barnard formed close attachments to older colleagues, always seeking a substitute father. Years after their work together at Lick, Barnard described in a letter to a friend his feeling for Burnham, saying "somehow my life seems wrapped up in his. Besides my most sincere admiration for him as a man and astronomer there seems to be a love for him that surmounts all else."

As much as the two respected one another, though, they had differing opinions regarding aspects of astronomy. One was on the issue of photography, which was fast becoming a valuable tool for astronomers. Burnham, even though he was an avid amateur photographer, preferred to remain strictly a visual astronomer. He expressed his opinion on astrophotography to Holden by saying, "[j]ust now there is a sort of craze on the subject, and a great many people will waste time which they might better employ." Barnard, on the other hand, was fully convinced of photography's power and was prepared to "waste time" exploiting it for his research. "There is no doubt but that [photography] is *the* Astronomy of the future," he wrote.

The ease with which Barnard accepted photography's role in astronomy resulted from having worked with it for over twenty years. Barnard had been born in 1857 in Nashville, Tennessee. He knew only poverty when he was a child, and had to drop out of grade school and take a job to help support himself and his mother. His mother gave him all his basic education. He found work in a photographer's studio, and through this job became expert in the techniques of photography. He knew how to adapt those techniques to work with telescopes when astronomy became his calling.

Barnard became acquainted with astronomy through books, and with the purchase of a 5-inch telescope (for $350, two-thirds his annual salary) he developed a passion for observing. He learned to hone his talents after he met Newcomb, the recognized grand dean of American astronomers, at a scientific meeting held in Nashville. Newcomb offered the eager young man advice on how to advance his astronomical work, and encouraged Barnard to acquire some of the mathematical skills denied him by his early departure from school. Barnard hired a tutor to help him with mathematics, but mostly it was hard work with the telescope that paid off for him. In 1881 he discovered his first comet, and he found several more in the following year. The discoveries earned him prize money, but more importantly they established his reputation for observing among astronomers and brought him to the attention of officials at Nashville's Vanderbilt University, who soon offered him a job.

Barnard took a position at Vanderbilt as part-time student, part-time instructor. It gave him the chance to take classes and fill out the analytical skills he needed to qualify as a professional astronomer, and at the same time to earn enough salary to support himself and his wife. Vanderbilt University owned a 6-inch refractor. Using it and his own 5-inch telescope, Barnard pushed on with his observing. He demonstrated his superb eyesight one evening while watching the moon occult a star. Since the moon has no atmosphere, the star's image should have blinked off instantly when the body of the moon crossed its position. Barnard noticed that when the star was occulted its image tarried dimly for a few tenths of a second, enough of a delay to bother him. He suspected the star might be a double, and that the delay was created as the two images were occulted in turn. His telescope, however, was too small to separate a close pair of stars. Burnham, working at Dearborn Observatory, subsequently confirmed his suspicion, and the discovery offered powerful testimony to Barnard's observing prowess.

Barnard was also looking ahead at this time, contacting people who could help his astronomical advancement. Holden was high on the list, especially after his appointment to the University of California presidency brought him closer to Lick Observatory. Barnard's many letters to Holden described his observations and contained sketches of uncatalogued (hence unknown and unstudied) nebulae. They also were richly embellished with unabashed praise for Holden, the note that any novice—especially one seeking a job—would strike with an established, eminent professional. Barnard once asked Holden to send a picture of himself, and then gushed thanks over the great gift when Holden complied. When Holden moved to California, his position at Lick essentially secured, Barnard said in a letter "I have said all along that you would be the man for the Lick Observatory. . . . [M]ay you live long to honor our land." Barnard saw in Holden a potential father figure, a thought he held until they finally met and he found that Holden did not match the figure he had imagined.

Barnard kept Holden aware of his experiments with celestial photography and his efforts to find the "chemical" (photographic) focus of Vanderbilt's telescope. Impressed by Barnard's growing reputation among astronomers, Holden decided to recruit the young man for Lick Observatory. Holden asked for a letter of resumé to give to the regents, and Barnard sent a note describing the ten comet discoveries he had made and the twenty-three new nebulae he had found. He also wrote, "I have made up my mind this summer to cease *comet* seeking and get into a higher class of work," in case Holden should think comet discoveries more show than science. He ended his letter "I am perfectly temperate, neither smoke, chew, nor use intoxicating drinks." Given the rough-and-tumble characters Floyd and Thomas Fraser had endured while building the observatory, Holden might have been gladdened by this claim to temperance. Sober or not, the excitable Barnard would cause Holden plenty of trouble in the years to come.

Barnard did not mention on his resumé that he was recklessly overeager. When Holden's offer of a position came, Barnard assumed it meant the observatory was ready for immediate operation. Before consulting with Holden about an appropriate arrival date, he quit his job and began selling his belongings and making plans to come west at full speed. When Holden wrote Barnard to delay his arrival, Barnard replied that he could not stop what he had begun, but that he would not mind observing in California without a salary for a while. What he did not yet understand was that Holden had no authority to assign him to work on the mountain, which was still under control of the Lick trustees. No salary for astronomers would be coming from the university until it possessed the observatory. When Barnard and his wife arrived late in the summer of 1887, they were forced to take a room in San Francisco for most of a year before they could move to Mount Hamilton. Barnard found a job in a law office to earn enough money to live on during the long wait.

Barnard's first chance to visit Mount Hamilton as other than a tourist came when the Lick trustees hired him to carry out the final inventory of the observatory. He named his price for the task as $25.00, plus expenses, but Floyd was able to offer a little better wages—$75.00 per month plus room and board. The job gave Barnard more than monetary reward, for it was his first chance to look through the telescopes, which he could do at night after counting and listing items during the day.

He was introduced to the telescopes by Keeler, a man about his own age but with a much different background in astronomy.

Keeler, unlike the self-taught Barnard and Burnham, had augmented his observing desires with university training in physics, especially in the techniques of spectroscopy. Holden had recommended to Floyd that Keeler be hired to help with the final preparations at the observatory, all the while planning to hire him later to head the spectroscopic work at Lick. Keeler and Holden had met several times over the years, starting when Holden was still at the Naval Observatory and Keeler was enrolled at Johns Hopkins University. They both took part in the 1878 eclipse expedition to Colorado, where Keeler put his artistic talents to work sketching the corona visible around the eclipsed sun. Keeler spent hours practicing his drawing by briefly scanning a scene and then sketching it from memory. His knack for holding details in mind until he could transfer them to paper was evident in his drawing of the eclipsed sun, which the other observers agreed was remarkably accurate. Holden included this drawing in a Naval Observatory report on the eclipse.

After graduating from Johns Hopkins, Keeler took a job with Samuel Langley, a self-taught scientist and one of the nation's first astrophysicists, at Allegheny Observatory, near Pittsburgh. Holden, then at Washburn Observatory, exchanged letters frequently with Langley, and through them heard nothing but praise for Keeler's work and abilities. And, like Barnard, Keeler made sure that Holden stayed aware of his research undertakings by sending him letters describing observations he had made and politely soliciting his opinions on matters of instrumentation. When Holden was appointed president of the University of California, and privately told Langley it meant he was assured of heading the observatory, Keeler wrote seeking a job at Lick. Holden quickly offered the young spectroscopist a post, but cautioned him that he could not officially hire anyone yet. Instead, Holden convinced the Lick trustees of the need for an astronomer on Mount Hamilton to help with the final preparations, and so Keeler came to California in 1886. He quickly became an indispensable aide to Floyd and Fraser, and also served as Holden's direct link with the mountain. Even though Holden was the prospective director, Floyd objected strenuously to his coming to Mount Hamilton, and Keeler had to act as his eyes and ears.

Besides establishing the observatory's time service and preparing the first volume of *Publications of Lick Observatory* for the trust, Keeler spent his own time exploring the skies above Mount Hamilton with the 12-inch refractor. His most important charge, however, was designing a spectroscope for the 36-inch telescope and planning the work to be done with it. The assignment involved probably the most scientifically important work the fledgling observatory would be undertaking, as it was Lick's entry into the new field of astrophysics. Keeler's background in physics, together with the light-gathering power of the 36-inch telescope, promised to make Lick Observatory a premier center for exploring the universe spectroscopically.

As much as telescopes, eyes, and photography allow astronomers to see what the universe holds, spectroscopy lets them more fully understand what they are viewing. As Keeler put it, "[T]he light which reveals to us the existence of the heavenly bodies also bears the secret of their constitution and physical condition." That secret is revealed by forming a spectrum, dispersing the light from an object into its underlying pattern of energies. A spectrum shows at what wavelengths, or colors, the light's energy is concentrated. This, in turn, provides the key for understanding the chemical make-up, physical state (pressure and temperature), and radial velocity of a star or nebula.

Photographic plates were still relatively insensitive when Keeler was learning his craft. Even stars bright enough to be recorded photographically would produce spectra too faint to register on photographic plates. With a star's light spread over a broad area, as in a spectrum, any point within that area is much fainter than the star itself. Most spectroscopic measurements, therefore, were made by eye on bright objects. Keeler had worked enough with Langley at Allegheny Observatory to know what the important elements of a well-designed spectroscope were from an observer's standpoint. His physics background, which included studies in optics, also enabled him to design the instrument on a solid theoretical basis.

Attached at the eye-end of the great refractor, the spectroscope took the light beam after it passed through its focus point and was diverging. The optical elements in the spectroscope had to realign the beam and direct it on either a prism or a finely ruled grating, which converted the beam into a spectrum. The design of the grating or prism determined how dispersed the beam became, which in turn set the spacing of the closest wavelengths that could be distinguished. After being dispersed into a spectrum, the image had to be focused by another lens so it could be examined. Keeler's spectroscope also included a micrometer for accurately measuring the separations of different spectral features.

The spectroscope was built by John Brashear, the Pennsylvanian optician and instrument maker who was Alvan Clark and Sons' chief rival. Brashear's shop was just a short distance from Allegheny Observatory, so Keeler had been a frequent visitor during his time there and had developed a close friendship with the older optician. The tubular frame supporting the spectroscope the two created extended a good three feet from the end of the already long telescope tube. It was built especially sturdily so that its flexure would not influence the observed positions of the spectral lines.

Holden considered Keeler's work, like Burnham's, sufficiently important to warrant two nights a week on the 36-inch telescope. At first Keeler made measurements that would enable him to determine the capabilities of the spectroscope and telescope combination so he could plot an effective research program. He observed some normal stars whose spectra were already known and could be compared to his sightings. Normal stars, like the sun, show energy over all wavelengths, but at some specific wavelengths there is a deficiency of energy, so-called absorption lines caused by the relatively cooler gas in a star's uppermost atmosphere absorbing energy radiated from the hotter, underlying body of the star. The gas atoms absorb at specific wavelengths as a result of their atomic structure. Keeler compared these observations to those of seemingly abnormal stars, whose spectra showed bright emission lines as well as the common absorption lines. Emission lines, bright dashes in the spectrum indicating more energy than the star normally would show at that wavelength, come from hot gas in extended envelopes or shells surrounding the stars. The abnormal stars Keeler observed are now known to be either expelling gas from their surface or to be part of binary systems whose members are exchanging gas, producing the unusual signature seen in the stars' spectra.

Keeler also observed several stars bright enough for him to measure the shift in their spectra created by their motion toward or away from us. These observations relied on what is known as the Doppler effect, in which emission from a star or other object will be shifted to bluer colors (shorter wavelengths) if the star is approaching, and toward the redder (longer wavelength) end of the spectrum if it is receding. The whole pattern of lines in a star's spectrum appears shifted, and by measuring the shifts of several lines—done by studying them visually and gauging their position relative to a benchmark—Keeler could calculate the star's radial motion. The reference wavelength came from an iron arc Keeler shined into the spectroscope. The micrometer enabled him to measure the positions of the spectral marks relative to the known wavelengths of the arc lamp's lines.

Holden considered measuring radial velocities of the stars the "principal spectroscopic work for which the great telescope was designed." These measurements would enable astronomers to determine the solar system's motion with respect to nearby stars, and would be the first step in mapping out the motions of stars in the Milky Way, a step that would lead to a better understanding of our home Galaxy's dynamics and distribution of stars. Keeler also thought that the program was important, but he was not as convinced as Holden of its practicality.

He considered it unlikely that the spectroscope could measure radial velocities for all the stars Holden wanted, because most were too faint for accurate work. Keeler thought it best to explore what the spectroscope was capable of before embarking on an extensive, long-term project.

To that end, and also because the work interested him more, Keeler began making spectral observations of nebulae such as Orion and assorted planetary nebulae. The spectra of these objects are characterized by bright emission lines arising from their gaseous nature. At the time, there was an ongoing debate regarding the width and exact wavelength of the strongest lines seen in the nebulae, which would in turn bear on the nebulae's composition. Keeler, aided by the powerful Lick refractor, the high dispersion provided by his spectroscope, and the sharp, steady images available from Mount Hamilton, was able to see these lines with greater resolution than others had achieved before. Ultimately he helped resolve the debate over the lines and lead the way to understanding the nebulae's nature. At first he was cautious about publishing his observations, because they were contrary to claims made by some of the most famous astronomers of his time. But as Keeler repeated his observations and got the same results his faith in them grew, and he wrote increasingly strong reports affirming his position, showing his confidence in the instrument and his ability to use it. Before long, astronomers at other institutions were able to confirm Keeler's observations, and they realized that he, not the earlier observers with their smaller telescopes, was right about the nebular lines. Keeler quickly made himself and Lick Observatory known in the national and international circles of astronomers.

As a college-educated bachelor, Keeler found a counterpart on Mount Hamilton in his colleague Schaeberle. The two lived in the two-story Brick House, as the astronomers' quarters were called, along with Holden and Burnham. Barnard and his wife lived in one of the cramped cottages that had been erected for the original work crew. The brick dormitory was split in half, with Holden occupying the apartment on one side and Burnham the one on the other. Burnham needed the extra space when his family came to visit, and Holden earned it by his position as director. Schaeberle and Keeler, meanwhile, each had only two rooms to themselves within the larger apartments. They lived separately, but organized a bachelor's mess to share food costs and meal preparations.

Even though their social situations were similar, the two differed greatly in their research interests. While Keeler was pushing hard at advancing the science of astrophysics, Schaeberle was happy to operate within the well-defined confines of classical astronomy. When Holden offered the young University of Michigan professor a position observing with the meridian circle, he replied, "I shall be particularly pleased to undertake fundamental work with the Meridian Circle; the creation of the Observatory is such that work of this kind can be executed under apparently peculiar advantages and conditions." Schaeberle's expression evinces his unexcitable style. Unlike Barnard, he had no plans to try for a "higher class" of work.

The positional measurements that Schaeberle made in steady, workmanlike fashion were important, but involved little scientific creativity. Working from a sliding chair—similar to an auto mechanic's cradle—beneath the eyepiece of the swinging meridian circle telescope, he sighted stars, timed their transits, and produced and refined long lists of star positions. Since the apparent positions of stars in the sky is affected by the atmosphere's refraction of light, Schaeberle devoted considerable time to determining the refraction as a function of the star's position above the horizon. His other major research interest was the corona of the sun. But as a quiet, retiring loner who adapted well to Mount Hamilton's isolation, Schaeberle was temperamentally well suited to the routines of fundamental astronomy.

Although all the staff members followed their separate research interests under Holden's general direction, there were some subjects that attracted and involved them all, mostly dealing with objects in the solar system. Mars, in particular, claimed great attention at the end of the nineteenth century because of published reports of observations hinting at the work of intelligent beings upon the red planet's surface. Most reports of these "canals," as the popular press misrepresented the original description of the orderly dark lines some observers had seen transecting Mars, were made with small telescopes. Naturally, the astronomers at Lick were eager to turn their large telescope and their own eyes to the question.

Through the 36-inch, Schaeberle claimed to see the canals, and on one evening he even described them as twin, parallel tracks, a sighting reported once before and one that would be difficult to explain as resulting from natural processes working on the planet's surface. Keeler, however, observing along with Schaeberle, recorded only faint, subtle shadings, not thin, linear features. Barnard, too, who had exceptionally keen eyesight, reported seeing only splotchy features.

Barnard thought that observers using small telescopes and working from sites without Mount Hamilton's steady atmosphere were fooled into seeing the boundaries between broad, irregular features as straight lines. The eye naturally seeks pattern and regularity, and if further motivated by a brain predisposed to believe in "canals," as in the case of Schaeberle, it is even more likely to see lines where an objective observer sees only divisions between faint shadings. Barnard was right. Space probes have since resolved the question of canals—there are none, although there was apparently once naturally flowing water on Mars.

Beyond Mars lay the giant gas planets, bodies many times bigger than the inner, rocky planets. Keeler, on the first night the 36-inch was used, had cast his gaze on Saturn and had produced a masterful drawing of the planet. Barnard, using the 12-inch, also studied Saturn, and observed an eclipse of one of its moons, Iapetus. When Iapetus passed behind the planet's rings, Barnard could see the moon shining faintly from behind the "crepe" ring, the innermost part of the ring system. Barnard's observation proved that this part of the planet's rings was at least partially transparent.

After his spectroscope was in operation, Keeler observed Saturn and its rings with it. He also trained the instrument on Jupiter, Uranus, and Neptune. In all the gas planets he saw broad absorption bands now known to be created by the molecule methane. At the time some astronomers had the idea that the outer planets might be self-luminous but Keeler's observations supported the explanation that they shine only by reflected sunlight.

Some of Keeler's more spectacular work did not involve the spectroscope, but only his eyes, artistic skills, and pencil and paper. After Saturn, he sketched Jupiter in many drawings that still stand as tributes to his artistic talent and perception of detail. From the ground, the eye is still one of the best instruments for studying the planets. During those brief moments when the atmosphere stands suddenly clear and unwavering, the planets' details jump out with absolute clarity. Before that image mentally fades, a fast and accurate hand can transfer it to paper.

Barnard was also competent at detailing through drawings what he observed. After he came to Lick he published a series of drawings he had made of Jupiter from 1879 to 1886 using his 5-inch telescope. He noted closely the interaction of Jupiter's Great Red Spot—presumably a huge cyclone traveling high in the planet's atmosphere—with the bands on either side of it. Not to be outdone, Holden published a similar article on Saturn, including in it his drawings and observations made with the telescopes at the Naval, Washburn, and Lick observatories. Holden's primary observing efforts from Mount Hamilton, however, were directed toward a more nearby body, the moon.

Holden reserved for his own work two nights each week on the 36-inch, which meant that he, Burnham, and Keeler took six nights with the telescope. The seventh night, Saturday, was used to show the heavens to members of the public willing to make the trip up the mountain. As with all of his bequests, James Lick considered the observatory a gift to the people of California. The evenings of public viewing, a tradition still carried on today, gave them a chance to enjoy that gift. The public showing ended around 11:00 P.M., and Burnham generally got use of the telescope for the remainder of the night. He only needed to change eyepieces and aim the telescope to commence observing. Keeler's spectroscope, on the other hand, took time to mount, and Holden's work required adding the photographic lens corrector to the telescope.

Holden was engaged in making a series of photographs of the moon at different phases for a lunar atlas that he intended to produce. He had first given Burnham this chore, but after he saw the dramatic pictures created by the telescope's long focal length and consequent large-scale images he decided to pursue the work himself. The director had Barnard develop most of the plates he took, relying on the young astronomer's years of experience in the Nashville photographer's shop to bring out carefully the details captured by the film emulsion. Barnard's critical eye noted that many of the images Holden took suffered from imperfect focusing or from vibrations, so that they were not as sharp as they could have been. Relegated to the 12-inch telescope for the first several years he was at Lick, Barnard hated to see the hours of 36-inch observing wasted on poor results.

Holden promoted his lunar photographs endlessly in reports to the university regents and in scientific articles. He described how Barnard had made positive images from the negatives captured on the photographic plates, and how a comparison of the two revealed details that might be missed by examining either alone. "[T]his method . . . is capable of producing new and important results, and of leading to veritable discoveries," he wrote. He went on so regularly and so effusively about his lunar studies that astronomers joked among themselves that Holden had discovered the moon. Back atop Mount Hamilton, Holden's colleagues, especially Barnard, thought the power of the 36-inch refractor might be applied to more significant work than merely taking pretty pictures of dubious scientific value.

Barnard got his chance to show his own skills at celestial photography using a 6-inch, wide-field lens. Known as the "Willard" lens, after the New York company that marketed it, the lens had been used for portrait work by a photographer in San Francisco. The large aperture of the lens greatly increased the amount of light that entered a camera and helped shorten portrait sittings during the days of slow-responding, wet-emulsion photographic plates. The owner had lent the lens to an amateur astronomer to photograph a solar eclipse, and the resulting images were so impressive that Holden purchased the lens ("at a very low price") for the observatory. Barnard experimented with it enough to show its potential for astronomy, after which Holden sent the lens to Brashear for some refiguring to optimize it for astronomical use. He then placed it in Barnard's charge. Barnard first made a camera out of a wooden box that held the lens and strapped on the side of the 6-inch refractor, but later the Willard lens gained its own equatorial mount and dome, and became known as the Crocker Photographic Telescope, after benefactor Charles F. Crocker, who paid to have the instrument built.

"[W]ith only the rudest equipment," Barnard began photographing comets that came into view. "It appears that there is now some hope of tracing [a comet] from day to day by photographic means," he wrote, "and of obtaining in this way some clue to the energy of the forces which produce [the] observed changes." Through a telescope Barnard saw a comet's tail as a faint, broad smear in the sky; his photographs showed the tail split into distinct streams of material, hinting at forces shaping the comet's cast-off material. Regularly photographing comets has developed into an essential part of the program to understand these ephemeral visitors to the inner solar system. Later, Barnard set an astronomical first when he discovered a comet photographically. Familiar with discovering comets by looking through a telescope at the night sky, it must have been a surprise for him to spot the comet's fuzzy streak on a plate taken through the Willard and developed in the dark room.

The power of a wide-field lens combined with a photographic plate was fully shown when Barnard turned his celestial camera toward the Milky Way, the broad swath of light created by the many millions of distant stars in our home Galaxy's disk. Barnard's pictures, "the only photographs ever made, here or elsewhere, that show at all the true Milky Way," as he characterized them, revealed "vast and wonderful cloud forms, with all their remarkable structure of lanes, holes and black gaps and sprays of stars." These impressive structures arise from the rich star fields embroidered with occluding swirls of gas and dust seen along the galactic disk.

Recording the Milky Way's appearance photographically required a delicate balance between the instrument's field of view and the exposure time. "In a photographic experience of twenty-five years," Barnard wrote, "I have never seen anything more deceptive to photograph than the Milky Way." His final success was the ultimate tribute to his skill, persistence, and dedication, for it required that he keep the telescope fixed precisely on the field, unwavering, for several hours, a duration guaranteed to strain eye, mind, and body. Following its exposure, the plate required "the utmost delicacy of treatment" to develop the faint features captured by the film emulsion.

Britain's Royal Astronomical Society later awarded Barnard its Gold Medal, the highest award of the Society, for his work on the Milky

Way. During the presentation, the speaker commented, "We must certainly admire not only the skill but the courage of a man who could, under the very shadow of the 36-inch refractor, demonstrate the merits of a lens which could be bought for a few shillings."

His photographic work was important for advancing astronomy, but Barnard's visual work went farther in capturing the headlines. Although he promised Holden he meant to graduate to a "higher class of work," Barnard was such a nonstop observer that he kept finding new comets, four in Lick's first year of operation alone. Holden regularly announced these discoveries to the local press to keep the observatory in the forefront of the public mind.

Time and again Barnard turned his gaze on Jupiter, first with the 12-inch telescope, and later with the great refractor. Working on the 12-inch, Barnard made studies of Jupiter's first satellite, today referred to as Io, as its orbital path carried it across the face of Jupiter. As the moon crossed Jupiter's light-colored bands Barnard saw it as two apparently separate, darkish blips. But when passing against a dark band on the planet, Io appeared to Barnard as an elongated white bar. From his first observations, Barnard wondered if the moon might actually be double, two moons circling together. As he continued his observations over more nights he realized that if Io were light in color along its equator and dark at the poles it would explain his observations. Modern studies have qualitatively confirmed Barnard's conclusion, but the fact that he could discern surface features on the distant, tiny moon indicates the high quality of his eyesight and his concentration at observing.

Circling Jupiter, another surprise lay in wait for Barnard, but one beyond the reach of the 12-inch telescope. For years Barnard had pressed Holden for time on the 36-inch telescope, and by appealing to the regents he finally was granted access to it in 1892. He wasted no time showing that he was qualified for the telescope, for only three months after he began observing with it he stunned the nation by discovering a fifth moon orbiting Jupiter. Galileo found the brightest four in 1610, when he trained the first telescope on Jupiter. Now, nearly three centuries later, Barnard stood beside his illustrious predecessor by finding a faint fifth member of the Jupiter family. Hundreds of astronomers had gazed on Jupiter over the years, but this discovery awaited a large enough telescope and a talented enough observer.

Holden sent news of the discovery by telegram to the Harvard College Observatory, which acted as a clearinghouse for astronomical discoveries. The Harvard director passed word of the finding on to the press, and the wire services carried the story back to California. Barnard became overnight a national celebrity, and was even more adored in the local areas. In subsequent years, several more of Jupiter's many moons were discovered by Lick astronomers, but none of these carried the excitement of Jupiter V, as Barnard's moon was called.

Barnard's impressive observational achievements in the first few years of Lick's operation caused Newcomb to remark later:

> A hundred astronomers might have used the appliances of the Lick Observatory for a whole generation without finding the fifth satellite of Jupiter; without successfully photographing the cloud forms of the Milky Way; without discovering the extraordinary patches of nebulous light, nearly or quite invisible to the naked eye, which fills some regions of the heavens.

Unlike Barnard's discoveries, the important spectroscopic work being done at Lick did not generate many headlines in the popular press. Begun by Keeler, the spectroscopic work was taken over in 1891 by his replacement William Wallace Campbell, the observatory's first new staff member since its opening. During the summer of 1890, Campbell had come to Mount Hamilton as a special volunteer, and had assisted Holden and Keeler with their observing. He became well acquainted with Keeler's spectroscope and the research program Holden envisioned. Campbell took an immediate liking to spectroscopic work, and learned all that Keeler could teach him during their few months together. At the time he probably could not have found a better teacher in the specialized science of astronomical spectroscopy. Although Campbell felt unready to work on his own when Holden's job offer came, Keeler encouraged him to accept the position, saying that only a handful of people in the country had more experience working with both a telescope and spectroscope.

Shortly after he hired Campbell, Holden got permission from the regents to add another astronomer, after having asked for many months for a larger staff to handle the workload. Holden settled on Henry Crew, another budding spectroscopist and the only member of the early staff to have an earned, not honorary, doctorate. Recognizing the importance of spectroscopy in pushing back the frontiers of astronomy, Holden thought having two spectroscopists on the staff would enable Lick to do more and better work in this field.

Crew had done his graduate work at Johns Hopkins University, and had studied there under Henry Rowland, recognized as the nation's expert in laboratory spectroscopic work. Rowland was masterful at designing new gratings and developing techniques to manufacture them, as well as using them for work in the laboratory identifying the wavelengths of resonance of various elements and molecules. His gratings could produce spectra with much higher dispersion than prisms could. Crew aimed to put to work at Lick Observatory the newest grating created by his former mentor, and planned his spectroscopic program accordingly. Unfortunately, many of the techniques he had learned in the laboratory were unsuitable when put into practice with a telescope, as he was soon to learn.

Holden expected Crew and Campbell to work together, but Crew balked at the forced collaboration. He cherished his independence so he could try his own research approach, and so he would not show Campbell his inexperience with a telescope. As a result of his tutoring by Keeler, Campbell was experienced with both the 36-inch and the only spectroscope on the mountain. Crew preferred to push along with his own research rather than become, essentially, Campbell's assistant with Keeler's instrument. Crew developed a spectroscope around a concave "Rowland grating," which had lines etched onto a curved surface rather than a flat one. The grating was designed to disperse light into a spectrum and focus it at the same time, thus reducing the number of optical elements in the spectroscope and using the light more efficiently. Crew believed that with the concave grating he could do more effective spectroscopy with the 36-inch.

Crew, however, had only worked with such a grating in the laboratory and with light sources much different than pinpoint star images. For star images, the advantages gained by using the grating are cancelled by its one major drawback—it not only disperses the light along wavelengths, but also spreads it perpendicular to wavelengths, widening the spectrum and weakening its surface brightness everywhere. In the end, the spectra that Crew produced at the telescope were too faint for any substantive analysis. The only data that he ever published from Lick were obtained with Keeler's spectroscope, not his own.

Campbell, meanwhile, was putting the 36-inch to good use nearly every night he was assigned to it. He improved Keeler's spectroscope by modifying it to use photographic plates, which with the development of dry emulsions were becoming sensitive enough to capture faint spectra. On plates, Campbell could record spectra of the same object on different nights and compare them for changes over time. And since the photographic plate could store up the image of a spectrum over a long exposure, Campbell could take spectra of objects too faint to study by eye. This meant he could embark on the program of measuring stellar radial velocities so favored by Holden.

Important scientific results were flowing from the young research center and the Lick staff was achieving recognition among scientists everywhere. But beneath the veneer of success there were steady rumblings of discontent. Frequent staff changes in the early years were signs of an underlying dissatisfaction that was eating away at Holden's command. Campbell was by this time the only Lick astronomer on good terms with the director. One by one the dissatisfied were leaving. Keeler, the first to go, had no hard feelings toward Holden, but felt that he needed to move away from Mount Hamilton to better his personal life. By 1891 he had been on the mountain for nearly five years and was tired of the limitations imposed by living at the isolated outpost. He was looking forward to marrying and starting a family, and Mount Hamilton, he felt, offered no suitable housing for that. After some searching and maneuvering he secured the director's position at Allegheny Observatory, which he had earlier left as an assistant to come to Lick.

Burnham, the next staff member to desert, returned to Chicago in 1892 for a court job that paid him over twice the $3,000 he was making annually at the observatory. Being able to spend more time with his family was certainly another consideration in his decision. The written blast he gave Holden as he left offered further evidence of his motivation for the move—he disliked the director immensely. Two months after Burnham's departure, Crew also left Lick. Like Burnham, Crew left when he landed a better job, but he too made his departure the opportunity for airing his hard feelings toward the director.

There were other significant departures in the observatory's early years—many of the people most intimately connected with the observatory's founding had died, closing the book on the beginnings of the mountain research station. Charles Feil, who had salvaged his family business from bankruptcy and produced the glass disks for the lens, had died in 1887 at the age of sixty-three. Later that year, eighty-year-old Alvan Clark, the crusty patriarch of the lens-making family, also died, shortly after he and his sons had finished figuring the 36-inch lens. Thus, the two people most responsible for fashioning the great refractor's objective lens were not among the living to hear of the successful first light on that cold winter night when Floyd, Swasey, Keeler, and Clark's son, Alvan G. Clark, first peered through the telescope.

For those near the project, however, the death that hit the hardest came in October 1890 when Floyd died. He had left Mount Hamilton several months before the University of California took over the observatory, and had never returned to see it fully staffed and operational. For a time, the prescribed rest at Kono Tayee and moderate quantities of claret helped, and he recovered somewhat. He even purchased a 5-inch lens from the Clarks (the business now carried on by the surviving sons) and built a crude telescope, with which he photographed the solar eclipse in 1889. He had hoped to devote more time to his home observatory, but his health took a turn for the worse. He went east to Philadelphia to seek help from heart specialists, but died there in a hospital, at the age of forty-seven. Cora Floyd, along with her daughter

Harry and niece Cora Matthews, had accompanied the Captain east. After his death she waited several weeks before she felt strong enough for the long ride back. Deeply upset by her husband's death, Cora was to die only four months later.

Floyd's funeral attracted the same group of San Franciscans as had come to Lick's earlier funeral. Members of the University of California Regents, the Society of California Pioneers, and the California Academy of Sciences, along with numerous city officials, were among the crowd. From the Lick staff, Keeler, Burnham, Schaeberle, and Barnard acted as pallbearers, but Holden conspicuously was not invited. Fraser traveled from southern California to San Francisco to attend. After he had left his job managing a ranch, Fraser had gone into business for himself in Banning, where he and his wife ran an inn. He also busied himself with some small farming. In 1891, a year after Floyd passed from the scene, Fraser died after a short illness. Like Floyd, Fraser lived only to the relatively young age of forty-seven.

Floyd and his associates had worked hard to build a durable monument to benefactor James Lick. They had constructed the sturdiest buildings they could, and had built a telescope intended for lasting contributions to astronomical research. But within Lick Observatory's walls there was steadily growing dissatisfaction among the staff. This was a source of strain the builders had not foreseen, and only time would tell how the observatory would respond to the increasing stress from within. The public started to wonder what was happening atop Mount Hamilton to make one respected staff member after another depart. They would discover soon, for the astronomers at Lick were headed for a showdown with their director that would play itself out at the observatory, at regents' meetings, in popular astronomy magazines, and most especially in the local press.

6

The Great I Am

1888–1897

By 1895 the battle lines on Mount Hamilton were clearly drawn. On one side was Director Edward S. Holden—imperious, dogmatic, holding on by force of his personality. Although he was almost without a supporter on the mountain, he still enjoyed the backing of the university regents. Opposing Holden were several independent-minded staff astronomers, their chief weapon their acid tongues. They were aided by the persuasive authority of newspapers and magazines. The disgruntled astronomers' tales of mismanagement by a pseudo-scientist on the mountain provided lively material for local reporters. Readers in the Bay Area began sharing the astronomers' views toward Holden as, one after another, respected members of the staff left Lick Observatory for other posts. Unfortunately for the beleaguered director, West Point had not offered training in the public relations battle being waged for control of Lick Observatory. With his support increasingly shaky, it seemed only a matter of time before the Czar of the mountain, as some stories characterized Holden, would take a big fall.

Several Lick astronomers disliked Holden immensely because of his habitual pose as an expert in nearly all aspects of astronomy, despite his inability to demonstrate more than a passing competence at actually doing research. His verbal outpourings led his opponents to refer to Holden as "the great I am," the man who had the answer for every question or situation. Another source of discontent between director and staff was the manner in which Holden exercised leadership. While at the Naval Observatory, which was directed by a naval officer, Holden had commented to Simon Newcomb that the place needed only stronger military-style discipline to become a truly first-rate institution. Even after leaving the Naval Observatory, Holden adhered to a military chain of command and expected his subordinates to carry out his orders without question. But the Lick astronomers, independent souls who disagreed with many of Holden's directions, did not believe in blind obedience. They might have followed Holden's orders out of respect for him, but he never worked to cultivate it.

Holden's attitudes kept him at odds with most of the Lick staff members, but many of his actions put him in a bad light with the university regents and the local populace. In the last weeks before the Lick Trust relinquished control of the observatory, Holden presented the trustees with his bill for $6,000 for services rendered. He claimed the sum as recompense for all the scientific advice he had provided them over the years. His bill listed the many letters he had written with suggestions for construction, the many trips he had taken on behalf of the trustees, the instruments he had ordered for them, and the contracts he had negotiated. Nowhere in Holden's account of the history of his involvement with the project did Newcomb's name appear, even though the new director's initial work for the trust had been in the role of Newcomb's assistant. Holden's bill was designed to make it seem as if he had been the sole inspiration behind the entire observatory.

The trustees were astounded, as was most of the public, when news of the bill from the "greedy astronomer," as Holden was immediately called, appeared in the papers. Richard S. Floyd was especially irked by the claim. He and the other trustees had worked to execute Lick's trust and build his observatory without ever receiving a salary for their efforts, so Holden's action was a personal affront to them. Working with pen from his sickbed at Kono Tayee, Floyd responded with his own account of the history of Holden's involvement. Floyd made it clear that he and the other trustees had considered Newcomb the principal scientific advisor to the Trust, and that Holden had insinuated himself into the project voluntarily. Newcomb also sent a letter describing his efforts with regard to the telescope and observatory. The trustees compiled other documents to accompany these reports in preparation for a legal battle over Holden's bill. When they deeded the observatory to the University of California and presented in gold the remainder of Lick's bequest, they held $10,000 in reserve in anticipation of the costs of a protracted court battle.

Holden's bill finally severed his relations with Floyd. Signs of potential friction between them had come even as far back as the negotiations with Warner and Swasey for building the telescope. Holden initially misled the midwestern engineers into believing that he was in charge of the telescope project, and attempted to reach an agreement with them without Floyd's consent. When the Captain found out, he quickly made clear that he was in charge of building the observatory. Later, when a design consideration came to Holden from the engineers, he sent it on to Floyd with the disclaimer that he could not care less how the matter was decided, a cavalier response that seemed to mock Floyd's assertion of control. Feelings of jealousy must have grown as Holden sat in his Berkeley office, denied by Floyd any access to the mountain observatory he was to direct. Holden's burning resentment of the Captain led him to claim that Floyd had gotten rid of Thomas Fraser near the end of the construction "for the purpose of being alone at the finish and of getting all the credit of the whole undertaking."

In the final months of construction, Holden tried to exert some control by continually asking Floyd to purchase books and assorted equipment that he regarded as necessities for the observatory. Floyd, who believed his own opinions were as good as any astronomer's, resisted to preserve some money for the observatory's operation. After becoming director, Holden continued to pressure the trustees to provide those items he felt were essential to the observatory's operation, and hence were the trustees' responsibility to furnish. He also complained about problems with the telescope's operation, and laid the fault with Floyd, who had not consulted him during the last year of construction. Holden described the observatory as a broken-down southern plantation, all show and no substance. He could not have chosen his words with better skill to inflame Floyd, a loyal southerner to his dying day.

Holden never took his case to court, and formally withdrew his claim in 1892, after which the trustees gave the final $10,000 to the regents. Even before that, however, Holden had quietly stopped pressing his bill. Publicly he said he had never really expected to be paid, but rather wanted only to establish for the record the value of his services to the trust. (His motive must have included putting into official records his version of the history of the observatory's construction.) Privately, however, he continued to state that on several occasions Floyd had promised to pay him for his services. The regents, who did not appreciate this controversial start to their new scientific institution, had pressured Holden to drop his claim. They felt that Holden's salary had already paid for any work relating to the observatory that he had performed while president of the University of California. He had earned $6,000 a year, and as observatory director he still drew $5,000, a handsome salary. The regents also feared that if Holden collected on his bill, they might be swamped with other claims from consultants across the country. Probably what finally made him desist, though, was the tremendously unfavorable press the affair was generating.

The newspapers had a field day with Holden's bill. Stories described the new director as "avaricious," "reaching," and "greedy." One head-line read "High Priced Star Gazer Meets Well Deserved Rebuff." The papers that had previously criticized the Lick trustees for squandering funds were now supporting them against Holden. This was the second time that Holden helped boost newspaper sales. The first had come while he still held the university's reins. Richard Proctor, an English author and lecturer, had written an article stating that no significant work could be expected from the Lick 36-inch telescope when it was finished because astronomers had never shown themselves able to apply skills learned with smaller telescopes to larger ones.

An editorial in a San Francisco paper labeled Proctor's ideas absurd. A reporter went to interview Holden for the "true facts." Though for the most part expressing himself in a reserved, diplomatic fashion, Holden ended the interview with some belittling remarks toward Proctor, which were reported as: "Mr. Proctor is not a working astronomer but a writer of books. . . . He has made the science familiar, but the telescope used by him was a small one, and he has never yet done anything of consequence. That is all I care to say for now, but a book would not hold all I could say." Proctor had thrown down the gauntlet, Holden had taken it up, and the battle was on.

Proctor responded to the report of the interview with a few fresh charges against Holden, as well as fantastic claims to his own important contributions to astronomy. As for Holden and his career, "[he] has done no single thing, either in the way of observation, thought or exposition, which will live beyond his own time, and he has done much which should make him thankful that such will be the fate of his work," Proctor said.

Holden, in turn, countered Proctor's charges and characterizations and added a few more digs at the English writer. This give-and-take found play on the pages of the *San Francisco Examiner,* but gained a wider audience when W. W. Payne, the editor of the popular astronomy magazine *Sidereal Messenger,* reprinted the exchanges. When he saw that Holden had gotten the final word in print, Proctor sent off a fresh reply to Payne. In true no-holds-barred fashion he exposed the tale of Holden's discovery of the supposed third moon of Mars, putting this mistake in public view for the first time. Holden, no doubt embarrassed by this low blow, wisely saw there would be no end to the exchanges if he continued to reply, and let the matter drop.

Astronomers generally and the university regents particularly did not like the controversy and were happy when it settled down. But local newspaper editors were glad to carry the story because it made such interesting reading. In spite of, or perhaps because of, his trial in the public's eye, Holden felt the observatory had an obligation to popularize its findings and spread the understanding of astronomy. To this end, in 1889 he founded the Astronomical Society of the Pacific, an organization that still prospers today. The society was composed of professional and amateur astronomers as well as people merely interested in astronomy. As founder and president of the society, as well as head of its Publications Committee, Holden's voice was the dominant one in the early *Publications of the Astronomical Society of the Pacific*. Through this medium he had a chance to build up his image and the observatory's, although at times his motives were so transparent that they worked against him.

In founding the society, Holden truly wanted to create a significant and lasting organization. He spent much time researching England's Royal Astronomical Society as a model for his new society to follow. He envisioned the society's goals as part public education and involvement, part stimulation for scientific inquiry. Astronomy was moving into a new era, confronting new problems, and Holden felt "the need for co-operation and concentration of forces is more and more pressing as the complexity of process becomes greater and greater." The Astronomical Society of the Pacific, Holden hoped, would enable astronomers to concentrate and combine their efforts. "There should be discussion, questions, remarks, interchange of ideas, contact of active minds. Let each member feel he has a part to bear, both in the actual meetings and outside of them, among his associates. In one word, let our society be a live one—active, intelligent, modest, competent."

Modesty left Holden when it fell to him to report in the society's publication on Lick Observatory activities. Here he told of recent discoveries by Lick astronomers, and dwelt at length on his own research efforts. He often presented the work of others in a way that hinted at some part in it by his own hand (or eye). In contrast, the discovery of Jupiter's fifth moon by E. E. Barnard, Holden's arch adversary on the Lick staff, received only a paragraph of coverage. Barnard, who enjoyed publicity himself, noted the sparse attention and commented that if Holden had made the discovery, the volume would not have been able to hold all the words the director would have written on the subject.

Founding the new society set Holden at odds with George Davidson, James Lick's original scientific advisor and a witness to the observatory's progress from its inception. Davidson presided over the California Academy of Sciences, and viewed it as the only legitimate scientific organization on the West Coast. It seemed to him that the Astronomical Society competed directly with the Academy, and Holden's position as Lick's director usurped his own claim to being California's premier astronomer. Davidson's jealousy of Holden made him an ally for the malcontented Lick astronomers in their conflict with the director.

In this sense, Holden's efforts with the Astronomical Society of the Pacific worked against him. Otherwise the association provided a bright point in his tenure. As a means of popularizing the science it was very successful. The Lick staff astronomers also gave talks to various school and social groups to share information about the work being done on the mountain and their own personal research interests. Lectures were a popular form of entertainment in the days before movies, radio, and television. Some of the astronomers, notably Barnard, whose talks were lively and entertaining, developed very devoted followings. Holden, too, was quite a personable speaker—critical newspaper stories notwithstanding—and his face was familiar to most school children in the area who learned about astronomy in their course work.

Holden achieved other benefits for Lick Observatory in its early days, particularly in the area of fund-raising. He was able to use his friendships with wealthy people in the Bay Area to persuade them to provide gifts to support the observatory's operation. The state of California, through the university, provided salaries and some annual operational costs, but Holden went before the regents again and again to report that the work at the observatory was hampered considerably by its small staff. He sought additional money to hire more astronomers. Unsuccessful there, Holden went so far as to take out notices in the newspapers soliciting gifts toward the observatory's operation, citing the large staffs at Harvard Observatory (forty employees), Paris (seventeen), and Pulkovo (sixteen) in contrast to Lick's tiny group of five astronomers. "[B]ig telescopes and fine instruments do not complete an observatory. . . . [I]t is apparent that five observers cannot do the work of twenty or thirty, and many questions of importance to the science must be neglected." To make good use of its many advantages, Holden wrote, Lick Observatory needed its friends to contribute. "[T]here is abundant chance for men of wealth who are interested in science to link their names with it by providing for its adequate support. . . . If California does not delay in giving this aid she can reap the glory of the discoveries to be made, and in the persons of her observers penetrate deeper into the mysteries of the skies than [can] other lands." Holden's powers of expression were never more eloquent than when he was appealing for money.

Though Holden's request for funds noted that men of wealth had a chance to link their name with science, one of his most enduring solicitations involved a wealthy woman, Catherine Wolfe Bruce, a reclusive New York spinster who had inherited a fortune from her father. In letters passed through Edward C. Pickering, director of the Harvard College Observatory and one of Bruce's few personal contacts, Holden laid out a plan to establish an award, the Bruce Medal, to be given by the Astronomical Society of the Pacific to the astronomer judged to have contributed most to the field over a specified span of time. The award recipient would be chosen by the directors of six prestigious observatories. Bruce provided $2,750 to establish a fund to support the award, after receiving assurances that it would go to foreign as well as American astronomers, and that it would be awarded only when there was a deserving candidate. The first recipient, in 1898, just before Holden left Lick Observatory, was Simon Newcomb.

Holden also managed to get Bruce to provide smaller amounts for other projects. He had similar success with a local heiress, Phoebe Apperson Hearst, the widow of Senator George Randolph Hearst and mother of William Randolph Hearst, of California publishing fame. She had a great interest in the University of California and served for a time on its Board of Regents, where her voice received considerable attention because of her wealth and generosity to the university. Holden was able to reap some of that generosity for the observatory's needs.

Hearst provided the funds for the first fellowship to support a graduate student at the observatory. The fellowship was short-lived, however, because in its second year Holden applied the money she contributed to finance an eclipse expedition. Hearst and other donors in the Bay Area also helped augment the observatory's equipment. From prominent banker and University of California Regent Charles F. Crocker, Holden obtained the money for the dome and telescope built around the Willard lens, which Barnard put to good use photographing comets and the Milky Way. Darius O. Mills, a member of the first Lick Trust and a prominent Californian, gave money for a new spectroscope for the 36-inch refractor. W. W. Campbell designed the spectroscope and put it into operation for the program of measuring stellar radial velocities. These sources also were good for many smaller gifts to the observatory, though their generosity rose and fell with the economy. At times these very wealthy people refused Holden's appeals, even though Lick Observatory's needs were modest.

Holden was ever resourceful, however, and even received from Thomas Edison's General Electric Company in New Jersey ". . . the complete plant of steam-engine and boiler, dynamo, belting, main wire, controlling wire, and a set of storage cells in duplicate, the whole as a free gift." With the addition of electricity, the observatory no longer had to rely exclusively on springs and gravity for power. Through the acquisition of such gifts, Holden performed a valuable service for the observatory, for he was able to further its research efforts substantially despite receiving only minimal support from the state legislature.

For the Lick staff astronomers, this grand effort was not enough to offset their dislike of the director. They did not appreciate Holden's desire to maintain tight supervision of their research. S. W. Burnham especially rebelled against it, for he had spent nearly his entire career working with his own telescope and at his own pace without the intervention of a director. Burnham was particularly upset when Holden criticized him for sending a research report to a journal for publication without first getting the director's approval of the report. This episode, which Burnham considered an attack on his scientific and personal integrity, came less than eight months after the observatory's opening. Burnham stayed at Lick for another three years, buoyed in part by his ironic sense of humor and his friendships with the other astronomers on the mountain. The easygoing astronomer was Holden's antithesis in many ways. Holden doted on his title and expected the staff to address him as "Director." Burnham disdained titles, and gave his definition of a "professor" as a person who "can swallow a sword 18 inches long, and eat glass out of a churn." Whether he meant that professors were strong-stomached or belonged in a circus sideshow is not clear, but it is obvious that he did not want the title for himself.

Finally, in 1892 Burnham accepted a higher-paying job as a court reporter back in Chicago. He let fly at Holden as he left, saying the director tried to rule the astronomers ". . . as though they were boys having no independent intelligence or self-respect of their own." Holden's military approach, in which he gave directions explicitly and severely criticized non-compliance, had finally taken its toll on the double-star observer. Several months later, Henry Crew followed him, airing his distaste for Holden in stories carried by the *San Jose Mercury* and the *San Francisco Examiner*. Crew had lasted at Lick Observatory for one year. He sent letters to his former professors giving his view of the situation at the observatory and hoping to receive backing for his actions, but instead they chided him for his disrespect to a superior.

These highly publicized resignations and departures were followed by articles in the papers describing the "misery" and "trouble" on Mount Hamilton, signed by "Observer" and "Astronomer." Though their anonymity was preserved, the authors, or at least the sources for the stories, were suspected to be Barnard and Burnham. Then an editorial appeared in *Astronomy and Astro-physics* recounting the departures of these respected astronomers and wondering if the director could do anything to save his staff and observatory. The forces of the press were molding public opinion strongly against Holden.

The only astronomers who appeared to side with Holden were Campbell and James E. Keeler. From his new position in Pittsburgh, Keeler sent Holden a letter saying that he could not have asked for fairer treatment while he worked at Lick. Holden quickly forwarded the letter to the university regents in an effort to retain their support. At Holden's suggestion, Keeler wrote in response to the *Astronomy and Astro-physics* article that he did not leave Lick because of mistreatment but rather for "private reasons which were in no way connected with the administration of observatory affairs. . . . I left Mount Hamilton with the good will of the Regents and on the best of terms with the Director and all my associates." Campbell, enjoying full freedom to carry out his spectroscopic work, also supported Holden and said the other astronomers really had no grounds for complaint. Holden had his faults, he admitted, but was a reasonable man if approached reasonably. In a letter to Keeler, Campbell said of Crew, "he is a bright fellow. But he had the 36-inch about 50 nights, and [obtained no] observations of the least value . . . to show for it." Campbell found a good deal of the fault for the situation to lie with the complainers themselves.

The sympathies of the other staff members at the observatory fell in between. John M. Schaeberle vacillated between supporting Holden and opposing him. On several occasions Schaeberle's support for Holden against the complaints of Barnard and others was crucial in persuading the regents that the director was acting responsibly. The younger staff members diplomatically and timidly tried to stay out of the arguments entirely. In 1892 Holden had hired Allen L. Colton to assist him with his program of lunar photography. The young astronomer quickly recognized Holden's lack of skill in astrophotography and believed that the director was deceitful in representing the work as mostly his own, since Colton knew that he himself had taken many of the best pictures. Although Colton was shy and fearful of speaking against the formidable director, he worked closely with Holden and could not help but see his faults.

The staff changes gave Holden other problems. For the emotional Barnard, Burnham's departure was a devastating blow; Burnham was his closest friend. But Barnard wasted no time asking Holden for the two nights on the 36-inch refractor freed by Burnham's leaving. "It is impossible for me to do anything like justice to myself and to the Lick Observatory without the use of the 36 in. equatorial. . . . Mr. Burnham's resignation leaves two nights a week at your disposal." Dispose of them Holden did, by assigning the nights to be shared between Schaeberle and Crew. Barnard had been requesting time on the telescope for years, and he did not take this latest denial lying down. He directed a letter to Timothy Guy Phelps, the chairman of the committee that reported to the regents on the observatory, complaining about his impasse with Holden over access to the great refractor. "In the hands of the Director," he wrote, "it has been a failure and his pure jealousy and spite will not permit me to have a chance. . . . I am sick of the continual efforts of the Director to crush me because I have opposed him in the Cause of right and Justice."

Barnard did not stop there. A second letter went to the full Board of Regents, outlining the situation as he saw it and asking them to consider the matter. He reported to them that on the few occasions he had gotten special access to the telescope, usually through Burnham's generosity, he had made significant discoveries, and regular use of it promised even more profound work. "For four years the telescope had been employed for a large part of the time in far less important work than that which I wish to put it to," he wrote, alluding to his estimation of the significance of Holden's lunar photography. To settle the matter, two members of the Lick Observatory Committee, Chairman Phelps and Regent Crocker, came to Mount Hamilton to interview the director and his staff. Based on their findings, the committee not only instructed Holden to give Barnard regular time on the 36-inch, but also recommended to the regents that they give the energetic young astronomer a substantial raise in salary. Barnard got his raise shortly thereafter, and then went on to discover Jupiter's fifth moon, fully justifying the committee's decision and refuting Holden's denials. On this round, at least, Barnard got the better of his nemesis.

Barnard and Holden's feuds ran back over many pages of correspondence, dating nearly from the time of the observatory's opening. Holden had a habit of communicating with his staff by leaving terse notes in their mailboxes. After Holden and Barnard started quarreling this soon became their only means of interaction. Barnard's messages were often extremely long-winded and detailed. These written exchanges document one long-running argument that developed after Barnard temporarily assumed the time-service duties when Keeler left on a two-month vacation. The day before Keeler's return an earthquake stopped all the observatory clocks and Barnard had to restart them. In doing so, he set them off by one minute, an error gross enough to be immediately noticed by the telegraphers along the Southern Pacific railroad line when the time signal was sent out. Holden became aware of the error and directed Barnard to reset the clocks. The next day a letter arrived from the Southern Pacific telegraph superintendent complaining of the error, and Holden showed the letter to Barnard to drive home the discredit his error had brought to Lick. Barnard responded to this reprimand with a written explanation of the circumstances in which he absolved himself of all culpability. In a sudden escalation reminiscent of Holden's fight with Proctor, the observatory files filled with accusations, blame, and excuses. Each accused the other of lying about how the situation began, and neither one trusted the other from that time on.

When Keeler left for Allegheny Observatory, Holden instructed Barnard to take over operation of the time service again. Barnard's response was that he was willing to handle it only temporarily, until Holden hired a new staff member who could assume the duties. To Barnard, it was obviously impossible to take on the job permanently because the important observational work in which the young rebel was engaged "... implies nights of sleepless labor, it means days of constant work ... to the neglect of health and other personal interests." Holden persisted and so did Barnard's negative responses: "I think it will be generally conceded that I observe *seven* nights out of the week, and these do not close at 10 or 12 o'clock, but continue throughout the entire night when the conditions are favorable." Holden, who had to stay fresh for his daytime administrative work and was an uninspired observer to begin with, only observed on his two scheduled nights on the 36-inch, and was notorious for cutting those nights short. To a fanatically dedicated astronomer like Barnard, such behavior was despicable.

In the end, Barnard won and did not have to take over the time service, which was assigned to Campbell when he arrived. On another issue, Barnard and Holden reached a stalemate. Barnard was eager to get his plates of the Milky Way printed for publication. To do the job properly, however, would be an expensive undertaking. Holden was quite willing to spend observatory funds to print his own moon atlas, but he found numerous excuses to put off printing Barnard's photographs, though astronomers generally recognized them as more significant scientifically.

This battle over the photographs continued for several years, until Barnard received an offer to come to the University of Chicago and work at its Yerkes Observatory, then under construction. George Ellery Hale, the driving force behind the new observatory, had secured funds from streetcar magnate Charles T. Yerkes to build the structure. It would feature a 40-inch refractor, a telescope destined to replace the Lick refractor as the world's largest. Based on their success with the 36-inch refractor, the company of Warner and Swasey was chosen to build the giant Yerkes refractor. (To this day, the Lick and Yerkes telescopes stand as the two largest refractors in the world.)

The Yerkes position appealed to Barnard strongly, but he hated the idea of leaving California's clear skies and mild climate for the cold midwestern winters. However, moving east would put him back in touch with Burnham, who had already accepted a post at Yerkes, and also free him from Holden. Barnard was so tightly wound that his constant conflict with Holden wore heavily on his health. His mental anxiety led to many physical ailments, some real, some imagined, as he himself recognized. While Barnard was agonizing over his decision, word of the offer leaked to the press, and thus became known to the regents. The prospect of losing their star astronomer dismayed them, and they hurried to ask Barnard if the rumors were true. Barnard denied that he was thinking of leaving, but in fact he had already considered it. Most knowledgeable observers suspected he would go. Holden was one. Knowing that he could outwait Barnard, Holden found new excuses to delay printing the photographs, even as Barnard redoubled his efforts to pry the money from observatory coffers. The photographs represented excellent scientific research done at Lick Observatory, and they should have been published under its auspices, but Holden saw a chance to save money and also to thwart Barnard.

To Holden's great relief, Barnard finally did accept the new position, and left Mount Hamilton in the fall of 1895. The astronomically minded citizens of California were greatly disappointed to see him go. The homespun young Southerner was a popular lecturer and his discovery of the fifth satellite of Jupiter had made him a local hero. His work had been rewarded by the astronomical community with such awards as the Lalande Prize of the French Academy of Sciences, and the Gold Medal of England's Royal Astronomical Society. Years later the Bruce Medal, which Holden had established, would go to Barnard: a final, ironic twist in the interplay between the two opponents.

The 40-inch Yerkes refractor, as well as enticing away the talented Barnard, seized Lick Observatory's claim to having the world's largest telescope. It is hard to imagine how James Lick might have responded to this turn of events. His great refractor had held the crown for only seven years. Holden, never idle, found a way to reestablish Lick's position in the number-one spot. Although the observatory could no longer boast the largest telescope, Holden had drawn up a plan for adding a 36-inch reflecting telescope to the mountain's arsenal. With California's steady skies and two 36-inch telescopes in operation, Lick would still justifiably be able to call itself the world's most powerful observatory.

Holden conceived his plan in 1893 when he read a notice that English amateur astronomer Edward Crossley wanted to "dispose of" a 3-foot reflecting telescope he owned. Now known as the "Crossley reflector," this telescope was designed and built by British engineer and astronomer Andrew Common. Common's countryman George Calver, who had initially proposed to the Lick trustees that they build a reflector, had cast the glass blank, which was shaped by Howard Grubb. The finished mirror was coated with a thin layer of chemically deposited silver. Common used the telescope for some valuable astronomical work. With it he demonstrated the power of astrophotography through a one-hour exposure of the Orion nebula. He subsequently won the Gold Medal of the Royal Astronomical Society for this work. Common sold this telescope to Crossley, a wealthy textile manufacturer, when he decided to build a larger, 5-foot reflector. Crossley's work with the telescope was bedeviled by the rainy, cloudy weather in Halifax. He built a large dome to shield the telescope, but by the time it was finished Crossley's "theological views had changed and no work of any kind was then to be thought of," according to a letter that Joseph Gledhill, Crossley's house astronomer, wrote to Holden explaining the telescope's availability.

From the beginning, both Holden and Floyd had felt that a large reflector was an essential complement to the 36-inch refractor. Now Holden leapt at the chance to achieve that goal. He wrote to Crossley asking for first rights to buy the instrument; Crossley replied, "There is nowhere I would sooner see the instrument than upon Mt. Hamilton." Holden quickly made an appeal to his usual list of prospective donors, seeking the money to buy the complete telescope and dome, which Crossley was offering at the modest price of $5,750. For a tiny fraction of what the Lick trustees had paid for the 36-inch refractor the observatory could acquire a large new telescope. Unfortunately for science, the wealthy San Franciscans Holden approached were more concerned with enduring an economic recession than realizing a bargain for the observatory. Unable to raise the funds, and knowing that Crossley had received an offer for the telescope from another source, Holden thought his dream had evaporated.

A year later, however, Holden's hopes revived. He received a letter from Gledhill telling him that the other sale had never materialized, and more importantly that Crossley might donate the telescope to Lick if Holden would request it. Holden's talents at persuasive communication stood him in good stead, and the artful letter he composed—plus Crossley's underlying generosity to science—won the donation of the telescope. Holden now needed only to find the money for dismantling and shipping the telescope and dome across the Atlantic and across the continent to the west coast. Lick Observatory's patrons in San Francisco were more easily swayed to part with smaller amounts of money. The many contributions totaled around $1,000, enough to cover the shipping. The donors included such luminaries as Levi Strauss, Charles F. Crocker, Darius O. Mills, and others, as duly noted on a marble plaque installed within the dome after its erection in California. Crossley demonstrated additional generosity, for he absorbed the extra cost himself when the transportation expenses came to more than he had first estimated. Holden further convinced the railroad companies to transport the telescope and dome across country free of charge. The University of California thus received a major new piece of equipment at no expense. The telescope and dome arrived on the mountain in the fall of 1895.

The Crossley telescope was proof of Holden's persuasive promotions of Lick Observatory's scientific importance and his abilities at fundraising. It put him momentarily back in good graces with the public. He made grandiose claims of the important findings that would come from the telescope. Almost simultaneously with the Crossley's arrival, Holden's harshest staff foe, Barnard, had departed for Yerkes Observatory. This should have been a time of great triumph for Holden, but instead the event sounded the first note in his final downfall.

As the Crossley was being shipped, Holden experienced a severe attack of sciatica that augured the hard times ahead for him. He spent several weeks in southern California, where he underwent an operation and recuperated. After a brief return to Mount Hamilton he traveled to the Naval Academy for a meeting of its board of visitors, his first trip east in a decade. Though he felt some attractions for the civilized life on the east coast (as opposed to Mount Hamilton's still rustic conditions), he regarded Lick Observatory and his position there as better than anything the eastern establishments could offer. At every opportunity he boasted to the scientists he met of Lick's superiority to other observatories.

When Holden returned to California, he found that the realities had not entirely changed after Barnard's departure. William J. Hussey, who had replaced Barnard, soon took up the same wrangles with Holden in which his predecessor had specialized. Worse yet, Campbell, one of the director's strongest remaining supporters, switched sides when Holden bungled a serious health problem on the mountain. Campbell's young son became very ill in the summer of 1896, and a doctor described his condition as acute malarial poisoning. The concerned father located the probable source of the disease in one of Mount Hamilton's water reservoirs, which he found littered with dead and decaying birds and rats. Campbell immediately sounded the alarm on the mountain and started boiling all his family's water. Holden reacted less vigorously. He sent water samples to the Berkeley campus for testing, but while awaiting the results did nothing to correct the problem. He did not believe the situation serious and thought Campbell was overreacting to it in hysteria over his son's illness. When the tests showed the water to be heavily contaminated with organic matter, Holden asked that the tests be run again and still did nothing to purify the mountain community's source of water. He began having fresh water hauled from Smith Creek for the mountain families, but let the public visitors continue to drink from the tainted reservoir supply.

This was the last straw for Campbell, who believed the director's inaction was a signal that he was "either monstrously incompetent or criminally negligent." As Holden continued to drag his feet, Campbell entered into the observatory records, where it would be available to the regents, his account of the episode. Holden countered with a report of his own account—pre-dated, according to Campbell—in which the director laid the blame for the affair on the young spectroscopist.

Holden's actions in this episode, as in other quarrels with his staff, typify a policy of burning his bridges behind him. His West Point training had given him the belief that a commanding officer could never admit a mistake, but in the world of astronomers his very denials aroused further antipathy. Aside from his moral obligation to the public health, Holden's long-term strategy would have been better served if he had kept Campbell as an ally. Instead, after this confrontation, Campbell joined the ranks of Holden's adversaries, dedicated to seeing him forced from the directorship. Holden might have worked harder to keep Campbell on his side if he had foreseen the rough times ahead with Hussey.

Hussey was another Michigan alumnus who had taken courses from Schaeberle and Campbell while an undergraduate. He had worked with Campbell at Ann Arbor after he got his degree, and then had come to California in 1892 as an assistant professor at newly opened Stanford University. He had hopes of getting a position at Lick Observatory when he came west. He worked there as a volunteer observer the summer he arrived, before he started teaching. Hussey and Holden began on good terms, mostly because Hussey recognized the value of having a friend in a high place and worked hard to ingratiate himself with the Lick director. His actions were effective, for Holden had Hussey in mind as a replacement for Barnard long before he actually departed. When Holden's offer of a job came, Hussey instantly accepted even though as a Lick astronomer his salary would be considerably lower than as a member of the Stanford faculty.

When Holden discussed with Hussey his potential work at Lick, he made clear that he expected the new staff member to take over Barnard's work with the 12-inch refractor and the wide-field photographic camera. He also promised him "certain nights with one of the two large telescopes," which could only have meant the Crossley reflector as well as the 36-inch refractor. The Crossley arrived on the mountain before Hussey did. Seeing the mass of metal parts, and realizing the chore their assembly would involve, Holden decided to postpone erecting the instrument until he could put Hussey in charge of the project. Hussey began working at Lick on New Year's Day, 1896, and the dome and telescope were largely in place by spring. Holden described the work as effected by "the Director assisted by Mr. Hussey." No doubt this meant that the observatory instrument technician did the lion's share of the work under Hussey's supervision, with Holden interjecting comments from the sidelines.

Merely assembling the dome and telescope was not enough to create a working astronomical instrument, however. The shakedown period for any new telescope is long, and for one as badly designed as the Crossley it seemed interminable. Hussey called the instrument a "pile of junk"; Barnard, in his parting shot at Holden, said he would not have paid $5.00 for it. Hussey wanted to observe, and although he was willing to spend some time trying to bring the Crossley into working condition he did not want to do that to the exclusion of research, for which he had come to Lick. No one else on the mountain was interested in giving up his own research, and Holden lacked the technical skill and time for the job. He asked Schaeberle to take on the project but the senior astronomer declined. Holden would not ask Campbell, because he considered the spectroscopist's work too valuable to interrupt. Although he had designed a spectrograph for the Crossley, Campbell had no time to make the telescope operational. With no other options, Holden pressed Hussey into the chore.

Like Barnard, Hussey had worked hard to cultivate Holden's favor, but then had taken an immediate dislike to him when he joined the Lick staff. Hussey had become a member of the Astronomical Society of the Pacific when he came to California, and shortly thereafter was elected the Society's president. Although Holden had stepped down from the presidency years before, he had kept his position of chairman of the Publications Committee, from which he could control the Society's journal. Hussey, from his new post, tried to remove Holden from the Publications Committee. He failed to muster the needed votes, but his attempt let Holden know that the new Lick staff member was now an enemy. Holden responded by telling Hussey to discontinue his observational work entirely and devote himself to his new assignment with the Crossley. Hussey, more composed than Barnard but equally willful, tried to block Holden's move. The other Lick astronomers supported his opposition, because they felt that if Holden could successfully reassign Hussey he could arbitrarily take them from their research just as easily.

Holden, however, was not proceeding completely arbitrarily. When he realized how long it would take to get the Crossley working he had approached the regents' Lick Observatory Committee about the problem. He spelled out for them his plan to turn the Crossley over to Hussey and the need for doing so. The committee supported his plan, so Hussey had no legitimate recourse when Holden gave him his instructions. Still, Hussey fought, countered, and resisted every order Holden made in the matter. He first demanded to see in writing Holden's authority to make the assignment. He would not follow any verbal commands from Holden, but waited to receive them in writing before he would respond. His protests of the situation included long letters that he demanded be copied into the observatory ledger and forwarded to the regents. Hussey did not intend to bow to what he saw as the director's caprice.

In the meantime, he made little progress with the Crossley, and all of 1896 drifted past. When the telescope arrived it really was a mechanic's nightmare and required considerable effort to make it operational. Common had designed a mount that did not place any large pieces of metal near the mirror, to avoid having air currents set up at night by the cooling metal that would interfere with the mirror's images. Thus, the framework for the telescope was a flimsy arrangement of thin metal members, making alignment of the main mirror and the upper, secondary mirror difficult. The mechanism for moving the telescope was extremely crude compared with the high standards set by Warner and Swasey for the 36-inch refractor. Preparing the Crossley reflector would have been a frustrating experience even for someone dedicated to making it operational. For Hussey, who despised the task, it was a never-ending bad dream.

Hussey worked on the Crossley, but every obstacle he encountered presented a new pretext for stopping. At each "impassable" difficulty, Hussey would write Holden a note saying he could not go further until the situation was fixed. Then he would sit back and wait. One day he discovered that the small flat mirror that reflected the converging light to its focus needed to be resilvered. He took it to Holden's office and left it with a note stating the problem. The director and astronomer butted heads loudly, Holden telling Hussey to resilver the flat, and Hussey responding that he would do so when he had the order in writing. Holden was livid. Schaeberle delicately resolved the standoff by resilvering the flat himself. Following that incident, Hussey found a problem with the drive clock, which makes the telescope track the stars. Holden sent the observatory's instrument technician to repair the clock. Without consulting the director, the upstart astronomer approved the technician's plea to take several days off before starting the work, and more time was lost.

Members of the Lick Observatory Committee again visited Mount Hamilton. They concluded that Holden was entirely in the right directing Hussey to prepare the Crossley for operation. Their visit finally sparked him to work on the telescope in earnest, and the Crossley, bad drive clock and all, was soon ready for some observing trials. Hussey tested the telescope visually on Saturn, and concluded that it was not even as good as the 12-inch refractor. He did not test the telescope photographically, as Holden had instructed him. The telescope's short focal length and the fact that it was a reflector, and thus by nature perfectly achromatic, made it much better suited for photography than for visual work.

Schaeberle, helping Hussey observe with the telescope on a subsequent night, noticed that the image was vibrating. He isolated the problem in the secondary mirror, which was improperly supported. Until the instrument worker, who had taken ill and was in a San Jose hospital, returned to Mount Hamilton to fix the secondary support, there was nothing more for Hussey to do. He applied for some vacation time. Holden denied Hussey's request, saying that the summer was the good observing season and that all staff astronomers were expected to be on the mountain working. But Hussey pointed out to the regents several past occasions when astronomers had taken time off in the summer, including Holden himself. The regents suggested to Holden that it might help ease tensions on the mountain if he acceded to Hussey's request.

The regents were closely watching this latest series of incidents, particularly as snippets about the struggle were appearing in the newspapers. They were eager to see the situation resolved without any additional criticism of the university or its observatory. Although the regents backed Holden in his arguments with Hussey, it was apparent that their general support for the director was beginning to slip. The regents cut his salary by $1,000 (they also cut the university president's by $2,000) in what they called an economy measure, but they surely would not have done so had they still thought highly of Holden. On his part, Holden was beginning to lose the spirit to keep up the fight. Since his bout with sciatica and trip to the east he had only rarely observed. His interests were shifting from astronomy to other pursuits. He worked on manuscripts and sent letters to his friends in the east, seeking a new position there. He told Newcomb that if he could find a job directing one of the large libraries in the east he would gladly leave Mount Hamilton. He began removing his pictures and diplomas from his office walls and quietly packing his belongings for what he told the staff would be a leave of absence.

Hussey, on his vacation from Mount Hamilton, continued his efforts to undermine Holden. He had help in this from several influential people. Hussey had maintained contact with his former boss, David Starr Jordan, the president of Stanford University. Hussey had stayed in touch with him to keep alive the chance of a return to Stanford should things not work out at Lick, but he also found Jordan to be an added source of support in his quarrel with Holden. "My suggestion is that you fight every point and not resign," the Stanford president wrote. "If you could rid California of the incubus of that immoral and incompetent man, it would be a public service of the first importance." Holden had been separated from his wife since coming to California, a serious breach of morality to the pious Jordan and other proper citizens of the era.

George Davidson also sided with Hussey, and the elderly astronomer could boast considerable influence in the San Francisco area. Davidson had tried to help Hussey topple Holden from the Astronomical Society of the Pacific's Publications Committee, even though he refused to join the rival organization himself. Working from the outside, he contacted all his friends among the amateur astronomers who were Society members and urged them to vote for Hussey's slate, but even his help was not enough to unseat the deeply entrenched Holden. Davidson continued to encourage Hussey, however, and tried to keep Holden on the defensive by sending damaging information to *Popular Astronomy*, the magazine edited by the gossip-loving Payne.

Hussey also found support among the university regents, a result of another of Holden's good ideas that had backfired. Holden had hoped to make Lick a center for graduate education, and in 1889 had accepted the first student, Armin O. Leuschner. Holden's suggestions for Leuschner's Ph.D. thesis never worked out and the student became increasingly disenchanted with the director. He gradually stopped work on his observational thesis, but found a niche at Berkeley, where he began teaching. He long held bad feelings toward Holden for what he considered the director's pretenses to competence in a field about which he knew little, and for steering him along a dead-end path in his initial research. In Berkeley, Leuschner met and eventually wed the daughter of Ernst A. Denicke, a strong force in the community and an *ex officio* member of the Board of Regents. Leuschner and Denicke shared a mutual dislike of Holden, which would ultimately count to Hussey's advantage.

Hussey wanted to hire a lawyer to attack Holden. Although Davidson tried to steer him in a different direction, Jordan thought the tactic was excellent. On Hussey's behalf he contacted E. L. Campbell, who had close connections with Stanford University and with its president. E. L. Campbell (no relation to astronomer W. W. Campbell) was a respected and normally high-priced attorney, but he took Hussey's case at no fee because he had decided, probably with Jordan's help, that "[i]n a matter of this kind the duty of discovery and exposure rests on every good citizen." Thus backed, Hussey used his vacation from Mount Hamilton to write a long, detailed account of conditions on the mountain, including many petty and small incidents. Part of his material came from rummages he had made through Holden's trash, or as he put it in his diary "chancing" upon a door he opened "out of curiosity" and "noticing" a paper that "fell" at his feet.

He had additional help in gathering damaging material against Holden. The observatory secretary, Charles D. Perrine, revealed confidential information to him. Perrine had grown increasingly disappointed when Holden continually passed him over when filling openings on the observing staff, and had become only too willing to help unseat the director. W. W. Campbell also kept Hussey apprised of Holden's latest actions, and gave him advice on composing his long complaint. Holden's quiet photographic assistant, Colton, also joined the cabal, sending Hussey notes describing Holden's past transgressions that he had witnessed.

Hussey prepared a document describing the Crossley telescope affair, his relation to it, and why he thought the regents had been wrong to side with Holden. He delivered this to the regents in Berkeley just before he returned to Mount Hamilton from his vacation, even though his lawyer did not think it could do much good. E. L. Campbell knew that the regents had acted properly in backing Holden, and that technically there was no recourse. He was at a loss for an opening for his legal attack against the director, and made plans to come to Mount Hamilton and interview astronomers Campbell, Perrine, and Colton for more material. Before the lawyer could do this, however, Colton bolted from the mountain. Whether he was completely fed up with Holden or emboldened by Hussey's example, he gave E. L. Campbell the opening he needed. The lawyer leapt to the task. The summer of 1897 suddenly became much hotter for the wilting director.

As he hastily left Mount Hamilton in August, Colton sent his resignation to the regents, along with a letter describing Holden's mistreatment of him personally and abuse of his scientific research. Holden recommended to the regents that Colton's resignation be accepted, and stated that he would be glad to give his side of the story whenever the regents cared to hear it. Holden also demonstrated that he had learned something about fighting a public relations battle, for through a friend in Oakland he fed derogatory reports on Colton to the newspapers.

After E. L. Campbell heard of Colton's resignation, he located him in San Francisco for several long interviews and plotted how to turn this event to his advantage. Colton was the source of several charges against Holden, which the lawyer pithily summarized as "Professional ignorance; Professional dishonesty; Administrative incapacity; Inveterate laziness and general unfitness to perform the duties of his office."

Through Colton, the lawyer also learned of Davidson's enmity for Holden, and through Davidson in turn he learned of a possible ally within the regents, Denicke. E. L. Campbell met with Denicke to sound him out on the best approach to take with the regents, and also to learn of other likely sympathizers. With Denicke's support, the lawyer requested the regents to examine the matter of Colton's resignation at their upcoming meeting in September.

The regents met on September 14, but the meeting dragged on and neared its end with no sign that Colton's letter would come up. Phelps, as chairman of the Lick Observatory Committee, had received the letter but was quietly holding it, hoping to avoid another embarrassing interchange involving Holden. Denicke knew how to force his hand, however, and rose and addressed the regents' chairman, Governor James H. Budd. Denicke demanded to know if he had seen Colton's letter of resignation. The governor admitted he had not, and steamily inquired why he had not been made aware of this matter. This quickly exposed Phelps' attempted cover-up. The hour was late, but under pressure from Denicke the regents agreed to reconvene on September 21 and take up Colton's resignation as their first item of business. E. L. Campbell knew he had won, because once he had had a chance to present the pile of damning material he had accumulated against Holden, the regents would not dare risk public wrath by further supporting the director.

But Holden made the point moot by leaving Mount Hamilton and Lick Observatory on September 18, 1897, the Saturday before the regents' meeting. He finished the process he had begun weeks before, packing his belongings and emptying his office and quarters. Brief notes in everyone's mailboxes constituted his final communication with the staff. His nine-year tenure as director was over, and his career as an astronomer was ended. His final dusty ride to town on the observatory stage, through the familiar sunlit hillsides dotted with oak groves, must have been a strange and somber one for the defeated scientist. The regents met as scheduled. There was some dispirited discussion of Holden, but even those who rose to defend him were tired of the situation. Despite strong pleas for support by Phelps, the regents nearly voted to call for Holden's resignation. Instead they selected an investigatory committee to report to them on the charges against Holden. Although E. L. Campbell knew that Holden had left the mountain, he did not want to allow any possibility of Holden returning to the directorship, and continued preparing his case for presentation to the regents. He made sure the public stayed involved by feeding copies of Colton's letter to reporters. E. L. Campbell now was acting as Colton's attorney, even though the quiet astronomer had already returned to Michigan.

The lawyer distilled the accounts of the Lick astronomers into a list of charges, which he showed to Phelps with the threat that he would officially present it to the regents and thus put it into the public domain unless the regents forced Holden's resignation. Holden had also departed from California by now, but he had left a letter of resignation with Phelps. When Phelps telegraphed Holden to notify him of the situation he said he had no choice but to offer the letter to the regents. The regents accepted the resignation unanimously, though they agreed to pay Holden through the end of the year. By then they planned to have hired a replacement.

Holden's letter of resignation, read at the regents' meeting, was predictably self-serving, recounting in exaggerated fashion his long involvement with the observatory. Clever to the end, Holden made use of his position on the Astronomical Society of the Pacific's Publications Committee, a post he still retained, to publish his letter in full in the Society's journal. Its appearance must have taken some of the sweetness from the taste of victory that Hussey and his accomplices were enjoying.

On the east coast, Holden tried to find a new position. He had hoped to become president of Massachusetts Institute of Technology, but Davidson and Burnham had friends at the school. Alerted by Holden's enemies, and perhaps also by the stories of the California controversy that had reached the eastern newspapers, the Institute's ruling body did not choose the aging West Pointer. He next sought the directorship of the Coast and Geodetic Survey, but again his reputation preceded him. At the Smithsonian Institution, now headed by Holden's old friend Samuel P. Langley, he also came away empty-handed. Simon Newcomb gave Holden strong recommendations for all these positions, but even the respected astronomer's backing was not enough now to overcome Holden's sudden fall from the heights.

Finally, the ex-director settled in New York and fell back on his writing talents, churning out books and magazine articles to make ends meet. He saw much more of his children than he had in California, and that provided him some solace. He continued supporting himself by his pen until 1901, when he was appointed to head the library at West Point. The ex-army officer had come home to stay. From then until his death in 1914, Holden taught cadets to find their way through the library, and shared with them his extensive insights into organizing, indexing, and researching.

Meanwhile, back in California, the regents faced a hard decision following Holden's departure. They still had a potent science center atop Mount Hamilton, but now it was rudderless, lacking a director to chart its course through the ocean of science. The regents assigned Schaeberle to take over the observatory temporarily, until a new director could be chosen. Given their difficulties with Lick's first director, the regents intended to take a long, hard look at the next person they appointed.

7

Adonais

1897–1900

Edward S. Holden left Lick Observatory in September 1897, but the University of California Regents agreed to make his resignation effective January 1, 1898. The grace period extended Holden's salary for a few months, a severance pay of sorts, and also gave the regents time to consider his successor. With the observatory off to a stormy beginning, they felt compelled to select a person respected in the local community, and someone not likely to stir up new controversy once in charge. The regents had two obvious choices: John M. Schaeberle, one of the original staff astronomers at Lick and now acting director of the observatory, and George Davidson, former president of the California Academy of Sciences. The regents were eager to make their decision quickly, so that the observatory could leave behind the sorry episode of Holden's forced departure. The vote on the new director was a major agenda item for the regents' meeting on December 14.

Schaeberle, a conservative astronomer dedicated to fundamental research, seemed likely if chosen to ease the strain atop Mount Hamilton and relieve pressure on the administration in Berkeley, too. He was well known in the area and had remained uninvolved during the various staff members' quarrels with Holden. As an astronomer, Schaeberle had continued to make positional measurements of the stars with the meridian circle, work he performed with competence but little imagination. He also studied the atmosphere's refraction of the sun's rays and starlight, and photographed eclipses with a long-focal length camera he had put together himself. Schaeberle was relaxed in his research activities, and was unlikely to set the world of astronomy aflame with bold or insightful work. His research was known to his contemporaries, but most of them did not consider it particularly exciting. A bachelor, Schaeberle had adapted well to the isolated life on Mount Hamilton, and seldom saw any need to travel to San Jose or San Francisco.

Davidson, Schaeberle's rival for the directorship, still considered himself the west coast's most eminent scientist. Although he looked upon himself as an astronomer first and foremost, his long career had been in geodesy—studying the shape of the earth. Now seventy-two years old, Davidson had maintained intimate connections with the observatory longer than anyone else, harking back to his conversations with James Lick about the project twenty-three years before. Some regents felt they owed Davidson a reward for this long-enduring association, and recompense for what he considered previous snubs. Davidson had been personally disappointed when Lick had reduced his original bequest for the observatory, and then had chosen Mount Hamilton rather than a peak in the Sierra as the observatory site. Furthermore, when Holden had come west, Davidson had viewed him as an interloper. During Holden's term as director, Davidson had stoked the dissension whenever he could, giving covert support to Burnham, Barnard, and Crew during their battles, and advising them of the continuing struggles after they had left Mount Hamilton. In this way he had helped keep up the criticism of Holden.

Davidson eagerly hoped for the Lick directorship, partly because he had lost his job with the Coast and Geodetic Survey two years before when he had turned seventy. He had gained many influential friends during his years in California, including several of the regents. His long tenure as president of the California Academy of Sciences had put him in contact with important civic leaders, on whose support he could also rely. On the other side, Schaeberle had the active backing of the Lick astronomers, who wrote the regents urging that the acting director be appointed as Holden's permanent successor. The Lick astronomers preferred Schaeberle to Davidson, whom they considered cantankerous and old-fashioned. They thought the aged geodesist lacked the scientific knowledge and experience to direct modern astronomical research. The December vote seemed likely to be a close one, and the two contestants quietly jockeyed for support.

But some of the regents decided they did not want to choose between just these two candidates when there was a world of astronomers to consider. Although Davidson was a sentimental favorite of the older regents, some of the newer regents did not share that enthusiasm. And though these more skeptical regents considered Schaeberle a possible candidate, they were concerned by his lukewarm reputation among the scientific community. University of California President Martin Kellogg wanted to increase the school's scientific strength, and he viewed the coming director's reputation and abilities as an important step toward this end. Kellogg was supported in this thinking by recently appointed regent Phoebe Apperson Hearst, a wealthy widow and the university's first female regent. Her generosity to the university—and to the observatory, before she had grown disenchanted with Holden—had been great and her fortune held the potential for many more gifts. Her opinion had considerable influence on the regents' thinking.

At the December 14 meeting the regents announced that they would postpone the decision, giving them time to seek more candidates. Schaeberle and Davidson must have been worried by this development. The newspapers eagerly joined in the speculation about the change in plans, and several new names cropped up in their columns, among them those of S. W. Burnham and E. E. Barnard. Davidson quietly alerted his two friends that they were being mentioned as candidates, and hence as his rivals. Each issued an announcement stating he was not interested in the job. Burnham went further and publicly lent his support to Davidson, but Barnard remained uncommitted.

Two other names came up in the press: one, Berkeley faculty member and long-time Holden foe Armin O. Leuschner, had no real chance for the post, but his father-in-law, Ernst A. Denicke, sat on the Board of Regents and that connection alone prompted speculation. The other name mentioned was that of James E. Keeler, ex-Lick astronomer and current director of Allegheny Observatory. The appearance of his name in the discussion indicated the increased respect he had gained among astronomers during his years at the underfunded Pennsylvania observatory.

The forty-year-old Keeler had traveled a long way since his childhood in La Salle, Illinois, on the banks of the Illinois River. Born in 1857, he was the second son of William F. Keeler and Anna Dutton Keeler. His father operated a foundry, but became a paymaster on the ironclad *Monitor* when the Civil War began. He guarded the coast for the Union while Richard S. Floyd was harassing shipping farther out at sea for the Confederacy. After the war, the Keeler family moved to Florida, where they found the winters milder than in Illinois. Rural Florida offered no proper schools for twelve-year-old Eddie, as his parents called him, so he learned from them at home. His father, who now operated a machine shop and forge, also taught him mechanical skills and instilled an aptitude for tinkering that would serve him well in his later career. Eddie avidly read *Scientific American* for information about the latest developments of the time. The magazine also carried advertisements for various scientific suppliers. Ed, then in his late teens, ordered a pair of lenses from one of them, and built a small telescope. He kept careful records of his observations, and made sketches of the moon and the planets. Even at this age his talents at drawing astronomical objects were evident.

Keeler's telescope helped pave the way for his formal education. His sister, attending a boarding school in New York, went with her class for a view through the telescope of wealthy amateur astronomer Charles H. Rockwell. When she mentioned that she had seen these same objects through her brother's telescope, Rockwell was intrigued enough to contact Keeler, and ultimately became his patron. Rockwell arranged for Keeler to come north for interviews with professors at Yale, Harvard, and the newly opened Johns Hopkins University, emphasizing to the school officials that Keeler's knowledge was self-gained and that his potential was great. The professors who met Keeler agreed, and thought Johns Hopkins was just the place for him.

Keeler had hoped instead for a job grinding lenses in Alvan Clark's optical shop, but fortunately for astronomy there were no openings there at the time. In 1877 Keeler joined the second class at Johns Hopkins. Daniel Coit Gilman, previously president at the University of California and now head of the new university, aimed to make it a respected center of scientific research. Gilman's approach enriched the undergraduates' studies with research-oriented learning. Since graduate students outnumbered undergraduates by two to one, the underclassmen were surrounded by students carrying out research. Keeler did well at school, and under the tutelage of Henry Rowland and Charles S. Hastings he learned valuable techniques in laboratory spectroscopy. Keeler majored in physics, and also took courses in German, mathematics, chemistry, and astronomy. Simon Newcomb, who lectured at Johns Hopkins part-time, taught the astronomy courses. By the time Keeler graduated in 1881, he had acquired a full complement of theoretical understanding to go with his nascent observational skills. He had no trouble finding work, and was hired by Samuel P. Langley at Allegheny Observatory. Two of Keeler's professors recommended him to Langley as an ideal assistant, one not blemished by an overly high estimation of his own knowledge, willing to admit his ignorance and work to fill in the gaps. While his classmates were receiving their diplomas in Baltimore, Keeler was already at work at the small observatory outside of Pittsburgh.

Allegheny Observatory had been founded in 1860 by a group of amateur astronomers, and featured a 13-inch refracting telescope. Langley had been appointed its director in 1867 after the observatory was taken over by the Western University of Pennsylvania, since renamed the University of Pittsburgh. A civil engineer by training and an architect by trade, Langley became a professional astronomer at the relatively late age of thirty. He began working as an assistant at the Harvard College Observatory, and quickly showed that he could adapt his analytical skills to this new field. When he took over at Allegheny he had been practicing astronomy for only three years, but had already made a name for himself. Unfortunately, Allegheny Observatory offered only the buildings, in a state of disrepair, and the telescope. It lacked auxiliary equipment and there was no monetary support. The practical Langley quickly conceived the scheme of measuring accurate time from the stars and selling subscriptions to this time service to the many railroad lines in the northeast industrial region. He also located a wealthy supporter, William Thaw, a Pittsburgh transportation magnate who provided money to support a variety of projects.

Langley was a true believer in the "new astronomy," as he called the developing field of astrophysics, and had pioneered many of the techniques in use by its practitioners. He had developed a bolometer, a detector containing a strip of platinum on which the light collected by the telescope was allowed to shine. As the platinum heated under this radiation its electrical resistance changed. Langley could measure this resistance and thus quantify the amount of incident radiation. These detectors, though remarkable for their time, were relatively insensitive in those early days, so most studies with them involved only the brightest astronomical objects, the sun and the moon.

Langley was interested in obtaining an accurate value for the sun's total radiation, which he knew from his studies at Allegheny was strongly absorbed by the earth's atmosphere. He planned an expedition to Mount Whitney, a 14,495-foot-high mountain in California, where he could measure the sun's light in the thin atmosphere. When Keeler arrived, he immediately went to work checking the instruments for the upcoming expedition, on which he would man the spectrobolometer, a bolometer coupled with a spectroscope to measure the sun's radiation at various wavelengths. Despite the difficulties of packing in their equipment through the rough terrain and conducting experiments at freezing heights, the expedition was very successful. Langley thought highly of the skills and resourcefulness Keeler demonstrated in making the measurements. Back at Allegheny, Keeler spent his time reliving his memories of the trip to the summit of the rugged mountain and reducing their observations.

The sun radiates a fair portion of its energy, it turned out, in the infrared where the human eye is not sensitive. For this work the bolometer was ideal, for from its response Keeler and Langley could tell that radiation was striking it even if they could not see the light. By manipulating their spectroscope and combining it with filters, they could study the bolometer's response and gauge the wavelengths at which the sun's infrared energy lay. Their study also indicated how the atmosphere's absorption affected the sun's apparent intensity at infrared wavelengths. Comparing measurements at Mount Whitney and at Allegheny, with their great differences in elevation, Langley learned that certain dominant absorption patterns in the sun's spectrum arose solely within the earth's atmosphere. They are known today to be due to water vapor.

Another important study came in 1882 when Venus transited the sun. Langley and Keeler looked for evidence of an atmosphere around Venus while it was backlit by the sun. They both did see a ring encircling the planet's disk, with a particularly bright spot on the ring. They independently sketched the planet's appearance and sent their drawings to one of Keeler's old professors, who studied their observations and concluded that the patch must arise from sunlight reflecting off a deep-lying cloud layer. This was an important step in understanding the atmosphere of Venus, and represented a fruitful collaboration between observer and theorist.

Even though Keeler continued his education at Allegheny—honing his observing ability, using instruments on the cutting edge of the science, and gaining a direct, personal appreciation for Langley's imaginative research approach—he was not satisfied with his position and longed to go back to school. Langley did not want to let his valuable assistant go, however, so they struck a bargain. Keeler agreed to stay at Allegheny for another year (even though it meant turning down a fellowship for graduate work at Johns Hopkins) and Langley in turn would persuade Thaw to finance Keeler for studies in Germany. As a result Keeler sailed for Europe in 1883, and enrolled at Heidelberg University and then in Berlin, the acknowledged centers of physical sciences and spectroscopy at the time. The years of hard study of German now paid off. Keeler got a first-class training in the theory of electromagnetic radiation (light), and a solid dose of optics along with it. Working in the laboratory as well as the classroom, Keeler undertook independent studies on the absorption of sunlight by carbon dioxide in the atmosphere, an extension of his work at Allegheny.

All too soon his time in Germany was up, and he returned to America and his job at Allegheny. By now he was too much of an independent thinker to be content as Langley's assistant. When Holden offered him the chance to develop the spectroscopic research program at Lick, Keeler jumped at it. Keeler made the most of his stay at Lick, and with the large refractor and a spectroscope of his own design he established himself as an important researcher. He also took steps toward improving his personal situation. Keeler fell in love with Cora Matthews, the niece of Cora Floyd and a frequent visitor to Mount Hamilton. Because of his close friendship with Captain Floyd, Keeler was often invited to family events, where he had many opportunities to be with Cora. His desire to marry her and the lack of a suitable house on the mountain for them to live in started him searching for a new post.

Keeler first told Timothy Guy Phelps, head of the Lick Observatory Committee, that his plans to marry meant he would need better housing if he were to stay at Lick. He hoped this hint would move the regents to build more housing. Since it seemed unlikely that they would do so, Keeler remained alert for other possibilities, and tried to manufacture some himself. His first attempt at a new position was at Stanford University, just getting started in 1890. Keeler laid a plan before one of the school's trustees for building a new astrophysical center that he would direct. Unfortunately, the trustees did not accept this plan. Keeler's next big push was at Allegheny Observatory, then an observatory adrift.

Shortly after Keeler had left for Lick, Langley had become the head of the Smithsonian Institution in Washington, D.C. Though he kept the title of director of Allegheny Observatory, he was hardly ever there and the telescope had essentially stood idle since his departure. Instrument maker John A. Brashear, whose shop was very near Allegheny Observatory, told Keeler that Langley might well be persuaded to resign the directorship, and asked if he would then be interested in the post. Keeler was, but did not intend to seek the position actively. He did not want to appear to undercut his former mentor. He wrote Brashear that if the trustees wanted him as director they should make him an offer.

Despite his statement to Brashear, Keeler did not remain idle. About this time he and Cora Matthews became officially engaged. When Keeler wrote her father asking for his consent, he said that he had already been offered the director's position at Allegheny. At the same time, Keeler prodded the university trustees in Pittsburgh, writing that the California regents were going to provide him with suitable quarters on Mount Hamilton if he would agree to stay at Lick. If the Allegheny trustees planned to offer him the directorship they had better act quickly, he implied. Actually, there was no such firm intention among the California regents, but Keeler was trying to galvanize the Allegheny trustees into action. He was able to pressure them further when Langley wrote and offered him a position at the Smithsonian, again as an assistant. Keeler used this new offer as another reason for the trustees of Western Pennsylvania University to make up their minds. Langley's offer, however, put Keeler in an awkward position; he did not want to tell Langley that he could not accept it because he was hoping to take over Allegheny Observatory itself.

His efforts might have come to nothing without Brashear, who was a dynamo of energy on Keeler's behalf. Brashear was eager to see Keeler return to Allegheny. He valued the younger man's friendship and greatly admired his talents. Brashear had a long connection with the observatory through the instruments he had made for it. He hated to see his carefully fashioned astronomical tools sit idle. Although he had no official capacity at the university, Brashear was well known to its officials. Through many personal contacts and persuasive arguments over a few hectic days he persuaded the trustees and the university's new chancellor to ease out Langley and hire Keeler to the post. Langley was initially upset, but quickly acquiesced to the decision. His disappointment resulted in part from the surprise of losing his by now honorary position, but most of all from realizing he would not have Keeler as his assistant again.

On receiving notice of his appointment, Keeler told Holden he was going to resign and telegraphed his acceptance to Pennsylvania. Through Phelps, he let the regents know of his plans and assured them he was leaving with the fondest feelings for Lick Observatory and the University of California. In June 1891, James and Cora left California for Pennsylvania, detouring to her parents' home in Louisiana for their wedding. Harry Floyd traveled with them. Although now legally an adult, Harry remained close to cousin Cora, who had been named as her guardian in Cora Floyd's will. The newlyweds and Harry continued east and settled into their new home in July, during the hot, muggy Pennsylvania summer.

Leaving Lick meant Keeler had given up working with the world's largest telescope. He had also traded Mount Hamilton's pristine skies for the smoky, sooty atmosphere of Pittsburgh, and left California's clear vistas for Pennsylvania's clouds. But Keeler was in charge at Allegheny. He could extend his research in any direction he desired, and his instruments permitted. He had gained virtually unlimited observing time with a 13-inch refractor. First, though, he had to put the observatory in order. It had been neglected for several years, and Keeler faced an extensive task just to get the dome open and the telescope working again. And, as Langley had when he first arrived, Keeler faced an observatory bare of auxiliary equipment. Langley had put into storage or taken with him all the optics and instruments he had acquired during his tenure. Langley felt that since Thaw had paid for them they properly belonged to his estate rather than to the observatory. In time, and with Brashear's help, Keeler gently persuaded Langley to return some of the equipment.

Fortunately for the new director, Thaw's widow and son continued the generosity to the observatory that the millionaire had begun. Besides paying the costs of the necessary work on the telescope and dome, their support enabled Keeler to order from Brashear a spectrograph for the 13-inch telescope. Keeler intended to continue the line of research he had begun at Lick, even though he knew he could not compete with W. W. Campbell, his successor at Lick. He had to modify his attack on account of the unsteady atmosphere and the relatively small telescope at Allegheny.

Keeler knew that by this time photographic spectroscopy had overtaken visual work, and he aimed to take full advantage of this new technology at Allegheny. To do so he had to teach himself photographic techniques. Keeler sought advice from as many experts as he could, and soon had built a professional-quality darkroom at the observatory. Even before he put his spectrograph into use, however, he realized a major shortcoming of the available photographic plates: they were least sensitive to the very wavelengths at which the telescope's lens and the spectrograph's prisms transmitted the most light. The most common types of plates were sensitive in the short-wavelength, blue region of the spectrum, but the lens of the Allegheny refractor absorbed much of the blue light and transmitted primarily the green, yellow, and red light.

Keeler realized he had to make efficient use of what light he had, especially since the relatively small aperture of the telescope was already restricting his science. He sought plates that were sensitive to the longer-wavelength parts of the spectrum, and learned sophisticated techniques of chemically enhancing the plates to increase their sensitivity. Keeler also developed a method of masking off parts of the photographic plate so that he could expose it to an object's spectrum, then shift the mask and expose a comparison spectrum on the same plate. He generally used the moon or the brightening dawn sky as the source of a spectrum of the sun for the comparison. This enabled him to measure the object's spectral features against known solar wavelengths. When Keeler finally had prepared himself and his instruments completely for photographic work, he began exploring a wavelength region that he had essentially all to himself.

Most of Keeler's work involved pinpointing and measuring whatever spectral lines an object emitted or absorbed, determining their exact wavelengths as accurately as he could. In this way, he and other spectroscopists of the day sought similarities and differences in the spectra of various objects in the sky. From his data Keeler drew what conclusions he could. As he had at Lick, and as Campbell was now doing there, he devoted much of his observing time to examining the Orion nebula and the stars within it. Keeler was able to show that several of the emission lines in the nebula occurred at the same wavelengths as absorption lines in the stars near the nebula, indicating that the stars and nebula contained the same material. This provided observational support to the idea that the stars had somehow formed from the nebular material.

At that time, laboratory spectroscopists were only beginning to learn which elements corresponded to which spectral lines. The interpretation of the spectra of nebulae was delayed for a long time because laboratory physicists could not duplicate the near-vacuum conditions that held in space; gas atoms were subjected to different conditions in the laboratory than in space. Theory had not yet caught up with astronomical observation, and for years the sources of many nebular spectral lines that were repeatedly seen remained unknown.

But for stars laboratory spectroscopy offered some help. Although Keeler knew that Campbell with his larger telescope could outdo him at routine stellar spectroscopy, he nevertheless spent part of his time observing stars. But he worked in a different spectral region than Campbell. By comparing the strengths of spectral lines he measured in stars with laboratory measurements of these same lines, Keeler was able to place the stars into a temperature sequence. Astronomers correctly assumed that stars condensed from nebular material. But they believed that after forming, stars gradually heated as they contracted, then slowly cooled after they had shrunk as much as they could. This implied the hottest stars could not be youngest. Keeler showed this simple picture would not account for his observational result that some of the stars in the Orion nebula, all presumably young, are among the hottest in the sky. His observation meant, as he stated, that all stars cannot be fitted into a single scheme of development. Nature is more complicated than that.

Even while he studied the stars and nebulae, the energetic observer did not ignore the planets. Keeler's most famous observation from his years as Allegheny's director, perhaps even in his career, was of Saturn and its rings. James Clerk Maxwell, who formalized the equations describing electromagnetic behavior, had predicted on theoretical grounds that Saturn's rings consisted of many small particles orbiting independently, like a stream of countless tiny moons. The alternative view, not accepted by serious astronomers but not disproved by observations, was that the rings were a solid disk.

Keeler knew that the two views could be distinguished by spectroscopy. If the rings were solid, the outer edge of the systems would rotate faster than the inner edge, just as the perimeter of a phonograph record moves faster than an interior groove (since it travels a longer circumference in the same period of time). Conversely, if the rings comprised individual particles orbiting under the control of Saturn's gravity, the outer ones would travel slower than the inner ones, just as Pluto, the most distant planet, orbits the sun slowest and Mercury, the innermost planet, orbits fastest. Keeler knew that the ring particles' velocities would reveal themselves through the Doppler shift, and he determined to get a good spectrogram of the rings.

Keeler had used this technique successfully to measure the rotation period of planets. As a planet spins, one edge is moving toward an observer and the other edge is moving away. Points along the planet's rotation axis, which forms a line in the plane of the sky that transects the planet's apparent disk, move perpendicular to the observer. Thus, from one side of the planet to the other there is a smooth progression of approach, tangential motion, and recession. The material at any point on the planet reflects a particular Doppler-shifted wavelength to Earth, depending on its motion toward Earth. Keeler could measure this effect by laying the slit of his spectrograph across the disk of a planet, perpendicular to the rotation axis. The spectral lines from the planet (the spectrum of the sunlight reflected from the planet) appeared tilted as the lines ran from their positions slightly blueward of their natural frequencies to their positions slightly redward. For fast-rotating planets such as Jupiter and Saturn the effect was fairly easy to measure, and Keeler had obtained accurate values for these planets' rotations.

For measuring Saturn's rings, the problem was that the sunlight reflecting from the countless tiny particles in the rings is much less than what reflects from the body of Saturn itself. When Keeler first got his spectrograph into operation, he tried to make the measurement but failed. The rings were too faint for his instrument in the blue spectral region (the sun shines brightest in yellow). When he obtained the yellow-sensitive photographic plates he was ready to try again, and in April 1895 he set up for a two-hour exposure on the rings. He placed the slit of the spectroscope across the planet and rings. In this manner one exposure would provide a spectrum of the rings on one side, from outer to inner perimeter, across the face of the planet, and of the rings on the other side in turn. When he finished the exposure he moved the mask he had built for his spectrograph and exposed the plate to the spectrum of the moon for comparison. Then he hurried into the darkroom.

When he developed the plate he saw what he had expected. The moon's spectral lines were vertical, for it rotates so slowly that its Doppler shift is negligible. The lines from Saturn and its rings were tilted. The planet's lines sloped appropriately for a rotating solid body, with the points on the outer edges having the highest radial velocity. The much fainter lines from the rings were canted also, but in the opposite direction, showing that their velocity was slowest at the outer edges, as expected for independently orbiting moonlets. The next night Keeler took a confirming plate. Within a few days he had measured the plates, and written up his results for publication. He sent his paper to the *Astrophysical Journal* for publication as quickly as he could, so that he would be the first to announce the observational confirmation of Maxwell's mathematical prediction. Keeler also called in a reporter to try to explain the story to him for local consumption.

Like Barnard's discovery of Jupiter's fifth moon, news of Keeler's observation traveled around the globe. It was important news to astronomers, and undoubtedly aided Keeler's fund-raising efforts in the Pittsburgh area. Shortly thereafter he was awarded the Rumford Medal, one of the highest prizes for American physicists, from the American Academy of Arts and Sciences for his work on Saturn's rings. It placed him in the company of such prior recipients of this prestigious medal as Thomas Alva Edison and Henry Rowland, his former professor at Johns Hopkins.

During his time at Allegheny, Keeler established a warm friendship with George Ellery Hale, who was located less than a day's train ride away in Chicago. The two had met when Hale had come to Lick on his honeymoon and seen Keeler working with the 36-inch refractor. Their friendship had been cemented on a trip to Pike's Peak, where together they tried to use an instrument Hale had designed to photograph the sun's corona without the benefit of an eclipse, but failed. At the time of Keeler's observation of Saturn's rings, Hale was bringing to completion the Yerkes Observatory and its 40-inch refractor. Hale was eager to get Keeler, by then considered the outstanding astronomical spectroscopist of his generation, to join the Yerkes staff. Had he done so, Keeler would have again worked beside Barnard and Burnham. Although he was attracted by the idea of going to Yerkes, Keeler felt that the Lick telescope, although surpassed in size, would remain superior because of its far better site. The 40-inch refractor was to be erected at Williams Bay, Wisconsin, where there are many fewer clear nights than at Mount Hamilton.

The University of Chicago, which administered Yerkes Observatory, was operating on a slender budget at the time. Despite all of Hale's considerable persuasive talents he could not convince the university president, William R. Harper, to offer Keeler a position. When Yerkes Observatory was dedicated in October 1897, Hale asked Keeler to deliver the main scientific address, giving Harper and the other university officials a chance to see him in action. The dedication was attended by many of America's most important astronomers, from the old guard, personified by the highly esteemed Simon Newcomb, to the forward-looking practitioners of the "new astronomy," such as Hale and Keeler. Keeler's address made clear where he thought the future lay. Astrophysics, as Keeler spelled it out, made use of ideas from every physical science that had something to offer—chemistry, physics, astronomy; it was not confined to orbital calculations or to simple positional measurements. Keeler outlined the importance of new techniques like photography and spectroscopy, and how they had helped develop the new science of astrophysics.

Keeler's address to the distinguished audience of scientists showed that he could be counted as a peer of the nation's most prestigious astronomers. Hale was still working strenuously to secure him a position at Yerkes. By this time, Keeler felt he had accomplished most of what he could hope to do at Allegheny Observatory, with its instrumental limitations. Because the Yerkes dedication took place just after Holden had departed from Lick, the meeting buzzed with speculation about his replacement. The tantalizing idea of becoming Lick's director surely ran through Keeler's thoughts, for Brashear had predicted years before that he would succeed Holden.

Even as Keeler was delivering his address at Yerkes, University of California President Kellogg was seeking names to add to the regents' list of candidates. One of the first new names suggested, by one of the Berkeley professors, was Simon Newcomb's. Newcomb had turned down the director's post when Lick trustees had approached him in 1874. Now, over twenty years later, the idea of ending his working career at Mount Hamilton appealed to him. Kellogg sought advice on both Newcomb and Davidson from Gilman, still the president at Johns Hopkins, and others. He learned that the general opinion was that while both were very respectable scientists, they were too old for the job, with their tempers not improved by age. The next director should be a vigorous man in the prime of his life, or so felt most of the administrators and astronomers Gilman had consulted. This supported Kellogg's own feelings, and he gladly passed the information on to the other regents.

Hale received a letter from Kellogg asking his opinion on Keeler as a candidate for the Lick directorship. Hale's thoughts must have been ambivalent, because on the one hand he had been pressing Harper to hire Keeler for Yerkes, but on the other hand he wanted his friend to have whatever was best for him. Hale contacted Keeler to see how he felt. Just as when the Allegheny post had become open six years before, Keeler replied that he would be interested in the Lick position if it were offered to him, but that he would not actively campaign for it. He especially did not want to create any bad blood between himself and Schaeberle, his former colleague. Hale wrote Kellogg to report that Keeler was probably interested in the directorship and to give him a strong endorsement, but he also warned the California president that he hoped to hire "the ablest spectroscopist in this country" to a position at Yerkes Observatory.

Unfortunately for Hale, the wheels that pry loose money turned slowly at the University of Chicago, while the University of California was pushing forward with its choice of a new director. In March 1898, the regents met again, and Kellogg put Keeler's name into consideration. Of the three candidates, Keeler had by far the strongest support from the progressive members of the scientific community. Keeler had moved ahead of Schaeberle and was fast overtaking Davidson, and it appeared that the regents would be badly divided. To avoid serious discord among the board members, one regent proposed a compromise in which the regents would elect Davidson as director, provided he would agree to step down within a few months and accept instead a professorship at Berkeley. In this way they would give him the recognition some thought he deserved, and a job, which he needed, but not at the expense of Lick Observatory's future. The plan offended Davidson's integrity, however, and he refused to accept it. He would either gain the directorship on his own strength or go down trying. When the regents voted they split evenly between Davidson and Keeler, and Schaeberle received only two votes. Following their standing procedure, they eliminated Schaeberle from the ballot and voted again. Keeler got the votes of both Schaeberle's supporters, and one of Davidson's supporters abstained. Keeler won.

The regents immediately telegraphed him and offered him the Lick directorship, but Keeler hesitated in accepting. Besides the still-beckoning Yerkes prospect, there was another possibility that had since developed. He and Brashear had generated some support among Pittsburgh businessmen for a plan to relocate the Allegheny Observatory to a more suitable site, and to build a larger telescope, a 30-inch refractor. Although their chances for raising the needed funds seemed promising, they had only a few pledges in hand when the Lick offer came. Still, Keeler had planted his roots in Pittsburgh, where his and Cora's two children had been born, and the prospects that were developing at Allegheny caused him to pause.

In the meantime, Hale had also been busy. He had secured from philanthropist Catherine Bruce enough funds to provide Keeler's salary for five years at Yerkes. Keeler also considered this offer, but because Harper would not guarantee his position or salary at the University of Chicago beyond the initial five years he would not accept it. With two small children to care for now job security was an important consideration, and he did not want to risk being unemployed five years later. It was not an easy decision for him to return to Lick, but it was undoubtedly the best for his astronomical career. After several weeks of deliberation Keeler accepted the Lick directorship.

Keeler enjoyed the backing of most the university regents, but his support among the Lick staff was by no means certain. They had endorsed Schaeberle in the regents' voting. Keeler knew Schaeberle and Campbell well, but the other senior astronomers—William J. Hussey, deeply involved in Holden's ouster, and Richard H. Tucker, a meridian-circle observer who worked with Schaeberle—had come to Lick after Keeler's departure. They did not know him, but knew of his respected spectroscopic research.

Schaeberle had difficulty in accepting Keeler's ascension. After the regents' decision he did not wish to stay longer at Lick. As the senior staff astronomer he was the natural choice for acting director, but if the regents did not want to confirm him as director, they did not want him at all he thought. Schaeberle stayed as acting director only until Keeler arrived, and then left Lick forever. Davidson, however, was a more gracious loser. He did not hold the regents' decision against Keeler, and wished him all good fortune as Lick's director. But he did harbor bitter sentiments toward the regents, who in his opinion had not selected the best candidate. His feelings were slightly mollified when the regents made him a professor of geography immediately after they had turned him down for the director's post. Joining the Berkeley faculty at least solved Davidson's unemployment problem.

For Keeler, the main question concerning relations with his staff centered on Campbell. His former protégé, too, must have felt some insecurity about Keeler's return because the two astronomers were working along very similar lines of research. Although Campbell had replaced Keeler's original instrument with a new spectrograph, the younger man still owed much of his expertise to the new director's tutoring. How would he now compare with his mentor? At separate institutions the two could remain friendly competitors, but at the same observatory with the same telescope and spectrograph they were all too likely to collide with each other trying to observe the same object. Campbell must have wondered if Keeler's return meant he would be relegated to a subordinate role in Lick's spectroscopic research.

As chance had it, Keeler arrived in California just as Campbell was coming back from an expedition to observe a solar eclipse. When Keeler and his wife arrived in San Francisco they were feted at a reception to meet many of the regents and university faculty members, and a few days later the scene was repeated in Berkeley. Keeler went up Mount Hamilton the following week, and found himself on the same stage with Campbell. One can imagine the two during the six-hour ride from San Jose, after exchanging greetings and sharing brief reminiscences, delicately exploring the situation created by Keeler's return. Perhaps it was then that Keeler made his decision, or perhaps he had already made it long before: to leave Campbell the 36-inch refractor and the spectroscopic work with it, and concentrate instead on getting the Crossley telescope into working order. When Keeler announced his intention he surprised and pleased the observatory staff.

The Crossley had stood idle since Hussey's halfhearted efforts to get it into operation. Keeler's astute sense of diplomacy told him that he would only generate more bad feelings if he assigned someone else to the Crossley. All the staff astronomers had strong negative attitudes towards the reflector, so the path of least resistance was for Keeler to take it on himself. But also, Keeler had come to realize from his experience at Allegheny that a refractor cheated him out of valuable parts of the spectrum, while a reflector would not. He hoped to solve the mystery of the faint nebulae, such as M 31, the Andromeda nebula, by obtaining their spectra. The Crossley reflector, working with a spectrograph of fast design, would be the ideal instrument for this program, Keeler knew.

When Keeler arrived at Allegheny he had faced the task of making its 13-inch telescope operational. It was a job for which he was well qualified, with his superb mechanical abilities and ten years of experience with telescopes and astronomical instruments. At Lick he faced a similarly difficult and time-consuming task, but now he had seven years more experience plus the support of a machinist, John McDonald, and an instrument maker, Emil Zengeler.

When the Crossley reflector had first arrived, Hussey had adjusted the telescope for the difference in latitude between Mount Hamilton and Halifax, where the telescope donor had his observatory. The Lick astronomer had not realized that the telescope had been designed and first used by Andrew Common in London, at a different latitude from Halifax. Hussey had wedged a correcting shim under the telescope's base but had never carefully checked the telescope's alignment to be sure that the polar axis pointed to the celestial north pole as viewed from Mount Hamilton. This error would have doomed any photographic work attempted with the telescope. With its polar axis not parallel to the earth's rotation axis, the telescope would have traced arcs across the sky different from the paths the stars followed, making long exposures impossible.

When Keeler inspected the Crossley he actually measured the inclination of the axis, and immediately noticed this problem. The telescope's alignment was incorrect by two degrees, four times the full moon's diameter, a gross error by astronomical standards. The error should have easily been apparent to Keeler's predecessors. The fact that it escaped their scrutiny demonstrates Hussey's profound indifference to the telescope and Holden's lack of direct involvement in the project and of critical scientific ability.

Keeler had McDonald, a holdover from Thomas Fraser's crew and the days of the observatory's construction, lower the telescope's pier by several feet, which increased the clearance between the top of the telescope and the interior of the dome roof. The telescope's 36-inch paraboloidal mirror reflected light back up and focused it at a point 17 feet above the lower end of the telescope. A small, tilted flat mirror brought the light to the side of the telescope, and an astronomer would either stand on a platform or mount a ladder to gain access to the image.

Next, under Keeler's direction, McDonald corrected the telescope's polar alignment by adding a concrete support base slanted at the proper angle. With graduate student Harold K. Palmer's assistance, Keeler spent many nights carefully aligning the mirror and the telescope mounting. Zengeler finally completed and installed the new clock drive, finishing the work he had begun for Hussey. Keeler and Palmer spent several more nights testing and adjusting the clock drive until it ran smoothly and made the telescope follow the stars' motions accurately. Keeler mounted and aligned finder telescopes for locating sky fields and guiding on celestial objects. He also removed the primary mirror's old silver coating and chemically deposited a new reflecting layer, employing a technique he had learned from Brashear. Finally, he renovated the darkroom in the dome building, for he intended to use the Crossley for astrophotography.

Even as Keeler was directing this task, and doing much of the work himself to carry it out, he had to perform all the additional administrative duties of the observatory director. He was responsible for the observatory property and its general upkeep. Like Holden before him, Keeler had to take care of the inventory, invoices, and orders for the observatory. All the papers had to go out over his signature. At Allegheny, where the staff consisted of Keeler and one assistant, the administrative tasks were fairly light, but at Lick, with its larger staff and remote location, the logistics were much more time consuming. Fortunately he had the assistance of Charles D. Perrine, the observatory secretary. As director, Keeler was also the focus of endless queries from the astronomically curious public. Answering them was a task that would try the patience of most astronomers. Keeler did not always enjoy this part of his job, but he met all questions with cheerful diligence and honest replies.

Keeler recognized that one of his responsibilities was to maintain a friendly atmosphere in which the staff astronomers could do their best research. Holden had given little thought to this, and under his administration, which had frequently run roughshod over his subordinates' feelings, the astronomers' thoughts were often far from their principal task, research. Keeler tried to keep Mount Hamilton quiet and peaceful, and the astronomers productive. One way he achieved this was to lead by example, showing the staff his own enthusiasm as a model. But he knew that a good attitude was only half the battle. From his own experience at Allegheny, where he had felt himself underpaid, he knew that a good salary does wonders in keeping up spirits. When he arrived at Lick, and Schaeberle had departed, Keeler had the opportunity of filling the vacated staff position. But instead Keeler decided to use the salary freed by the opening to fund fellowships so that several graduate students could come to Mount Hamilton to do research. His plan left enough money for raises for all the staff astronomers, a gesture that solidified their good feelings toward Keeler.

He also worked to improve relations between the Mount Hamilton staff and the Berkeley faculty, which had suffered under Holden's antagonistic brand of independence. Keeler arranged for the Lick astronomers to go to Berkeley and teach courses periodically, a move that fostered cooperation between them and the professors on the campus. It also provided the undergraduates with firsthand lessons from practicing astronomers. This marked the beginning of a working collaboration between the observatory and the Berkeley astronomy department that lasted for many years.

For his own chief task, Keeler got the Crossley reflector into working order and began doing research with it. He realized that because of the small f-ratio and relatively large field of view the telescope would be able to take photographs better than any that astronomers had previously achieved. He had an instrument with tremendous light-gathering power and the pristine Mount Hamilton skies under which to work. He was eager to get going. Keeler arrived on Mount Hamilton on June 1, 1898, ten years to the day after the observatory was officially dedicated, and by November he was taking his first photographs with the Crossley. He had accomplished in five months what his predecessors had failed to achieve in three years.

Keeler's first direct photographs were of Comet Brooks, a passing fancy for him that nonetheless gave the Crossley a good workout. Tracking the comet let Keeler check the telescope's performance under actual observing conditions. He guided the telescope with a long-focus finder to stay fixed on the moving comet. For exposures on star fields Keeler had devised a photographic plate holder that had adjusting screws on two sides. Manipulating the screws shifted the plate and kept the field of sky precisely positioned on it. Despite Keeler's improvements to the telescope, it was still an awkward mechanical instrument. Imperfections in its gearing guaranteed an imperfect match to the sky's motion while it tracked. Keeler's plate holder effectively solved this problem.

Studying Comet Brooks also let Keeler experiment with the Crossley's photographic abilities. The telescope produced some splendid pictures, detailed close-ups of the comet's head unlike the wide views provided by cameras such as the Willard lens that Barnard had used. The reflector was a success, but this trial run suggested to Keeler some additional needed refinements. After these touch-ups he aimed the reflector at what he considered a real research object, the Orion nebula, an object that held great interest for both him and Campbell. Keeler had studied the Orion nebula at Allegheny. Likewise, Campbell had observed it from Mount Hamilton, where he found that the strength of the two brightest green nebular lines varied across the nebula. Generally, Campbell saw the lines becoming fainter with respect to the bright hydrogen line known as H-beta as he proceeded out from the nebula's central bright stars. When Keeler arrived at Mount Hamilton he had joined Campbell at the 36-inch refractor and confirmed this observation. Campbell's idea put him at odds with some of the most respected German spectroscopists, who suggested that the variation was not real, but was instead due to the eye's changing response. They were led to this conclusion because they could not verify the observation with their telescopes, limited in size and located under poor skies. Using the Crossley, Keeler attempted to resolve the controversy.

Keeler had no spectrograph (Campbell had had one made, but it was poorly designed, poorly built, and never used). Instead, he acquired several filters with which he could isolate particular spectral regions, each passing only the light within a particular range of wavelengths. Taking direct photographs through the filters, Keeler could record the nebula's appearance and brightness in different spectral lines. Keeler's photographs demonstrated that the brightness of the lines in the nebula did vary as Campbell had reported and as Keeler had confirmed visually. And, because photographs do not lie (although Keeler was always suspicious of anything faintly seen on one plate but not reproduced on another photograph of the same object), Campbell's opponents had to accept that they were wrong and that he was right.

This important study was mere testing of the waters for Keeler. His most significant work with the Crossley was on the spiral nebulae, very different objects from the nebula in Orion. Several of these odd-looking spiral nebulae could be seen visually with telescopes, and had been studied previously. Their spiral appearance, suggestive of a cosmic whirlpool, seemed to indicate that they were rotating. Astronomers generally believed that in the spiral nebulae they were witnessing new stars and planetary systems condensing from gaseous clouds in space.

Keeler observed the spiral nebulae with enthusiasm. His fine-tuning of the Crossley allowed him to expose his photographic plates accurately for long times, up to four hours in some cases. These long exposures plus the Crossley's 3-foot mirror, which captured copious amounts of light, enabled Keeler to record faint details never before seen in these objects, "arches of nebulosity" or "long thin branches. . . . and the spiral turns at their extremeties." Significantly, Keeler's long-exposure photographs of the larger spirals revealed numerous smaller, fainter nebulae around them. Some were obviously miniature versions of spirals, distinguished by their characteristic swirl. Others appeared as indistinct roundish blobs. Still others looked like short bright lines bulged at their centers. Keeler realized that if the spiral nebulae were in fact flattened objects, many of these smaller objects he had captured could be identified as spirals tilted at various angles relative to the line of sight. The lines with bulged centers he recognized as spirals viewed edge-on; he could even see a dark lane in one crossing it lengthwise. Had he considered this further he might have connected it with the "black cracks or crevices" Barnard had photographed tracing the Milky Way, and thus skipped ahead several decades in astronomical thinking.

As it was, Keeler recognized that the spiral systems were flattened and that there were many of these objects in the universe, the majority of them too small and faint to be recorded by most existing telescopes. He estimated that there must be 120,000 such nebulae within reach of the Crossley. Keeler reckoned that they far outnumbered the other nebulae, such as planetary nebulae or ones like Orion, and so must be an important constituent of the universe. Nearly every photograph he took turned up new nebulae to be added to the list of objects for study.

Today's astronomers recognize these nebulae as galaxies of stars in their own right, similar to our own home Galaxy but incredibly far away, and indeed, as Keeler had suggested, objects ubiquitous in the universe. But in Keeler's time that view was still far in the future, and he was providing the first real information that the nebulae existed in such large numbers. His photographs were offering the first solid clues to their structure.

While Keeler was occupied with this frontier research, in the Midwest Hale was working to establish the Astronomical and Astrophysical Society, an organization that has since evolved into the nation's largest professional astronomical organization. (It is now called the American Astronomical Society.) A meeting of the fledgling society convened at Yerkes Observatory in 1899, and Keeler sent copies of some of his Crossley photographs there for display. The astronomers in attendance were deeply impressed by what they saw. Barnard, whose observing skills were unparalleled, marveled at the pictures for hours, exclaiming over the detail they presented. His response on this occasion contrasts with his public statement when he left Lick Observatory that he would not have paid $5.00 for the Crossley telescope. Hale himself remarked that the photographs "created a genuine sensation and showed to many who had been skeptical regarding the advantages of reflectors what the instrument is capable of doing in the right hands."

Holden's skills of persuasion had acquired the Crossley reflector for the observatory, but it took Keeler's willingness, dedication, and expertise to make a working instrument of it. In doing so, he established for the Crossley a historic position in the annals of telescope development. For many years it was a workhorse for research at Lick Observatory. Keeler's work pushed reflectors to the forefront of astronomical observing, and all major research telescopes built since his time have been reflectors. It seemed that for Keeler every scientific problem he touched produced golden results. Throughout his career he made his mark on astronomy. In the tradition of Langley and Rowland, Keeler was a superb instrumentalist. One of the most difficult tasks for an astronomer is to design an instrument, such as a spectrograph, to work well with a particular telescope and also be suited for a planned program of research. Keeler's intelligence, training, and experience combined to make him an expert at this rare skill.

Keeler's scientific generosity was also often evidenced during his career. Though he made sure that he received proper credit for the work he did, he did not hold back the advancement of astronomy from a selfish desire to do a particular research program himself. For example, with the Crossley he had made some exploratory photographic searches of globular clusters: compact, symmetric associations of many thousands of stars. At the same time, Solon I. Bailey, an astronomer at Harvard, had begun a photometric study of variable stars within globular clusters, but his work was hampered by the small telescope he had. Edward C. Pickering, director of Harvard College Observatory, asked for Keeler's help with the project and the Lick director complied, taking a series of exposures of various clusters. Bailey had the skills for making quantitative photometric measurements from photographic plates, and Keeler had the instrument and observing talents for accumulating the data. Their combined efforts formed a successful, cross-country collaboration, and demonstrated Keeler's willingness to set aside his own personal research briefly to help with a worthy project.

Keeler had truly come home when he returned to Mount Hamilton. Through his research at Lick Observatory he solidified the reputation he had gained among the world's astronomers. In a very short time he had set the observatory back on its feet so that it was again a productive and harmonious research center. And he had put a major new astronomical instrument into operation. Still a young man with much creativity to offer, he was looking forward to a long career, but instead he was struck down in the prime of his life. Through the winter of 1899 and on into the spring of 1900, Keeler suffered from what he thought was a bad cold, one he could not shake. He grew increasingly weaker with the passing months until any exertion would leave him short of breath and flushed in the face. Soon, he could not even make the short climb from the Crossley dome, on Ptolemy Ridge below the summit, to the Main Building. He quietly prepared a list of his biographical data and left it on file at the observatory, as if he knew that someone would soon be writing his obituary.

Keeler spent a vacation at the seashore with his family over the fourth of July, and then returned to Mount Hamilton under his doctor's orders not to do any observing. He spent the rest of July at Mount Hamilton, but at the end of the month he went down to visit his doctor again, then headed for Kono Tayee where he met his family, and planned to recuperate for several weeks. His plans did not work out and his condition worsened. Keeler was traveling back to San Jose with Cora to see his physician again when he suffered a stroke. He stayed in San Francisco that night, under the care of the city's best doctors. Two days later, on August 12, 1900, he suffered a second stroke and died, one month before his forty-third birthday.

After Keeler's body was cremated Campbell proposed depositing the ashes in the vault with Lick's corpse at the base of the 36-inch refractor (perhaps the Crossley would have been the more appropriate instrument). Cora Keeler favored this plan, but the regents rejected it, stating that the great refractor stood for Lick alone. Campbell then proposed another spot for Keeler's ashes on Mount Hamilton, but Cora was too shaken to consider the matter further. They were finally entombed in the dome of the Keeler reflector at Allegheny Observatory, a telescope made possible by Keeler's initial fund-raising exertions and pushed to completion by his good friend Brashear.

The astronomers at Lick and across the nation were stunned by Keeler's death in the light of the powerful contributions he had made to science, and the even more powerful ones they expected him to make in the balance of his career. But beyond that, there was deep respect and admiration for Keeler as a person. "My principle is, do the best you can, and then laugh if you fail," he had once written to his friend Hale. This attitude pervaded his approach to science, and though he took astronomy very seriously he did not let science dominate his life. Keeler possessed a deep sense of justice. When Holden had directed the observatory, he had done all he could to diminish Floyd's contribution to building the observatory and instead claim credit himself. When Keeler arrived, one of his first actions had been to mount a large portrait of Floyd in a prominent position in the Main Building, restoring Floyd to a place of honor. His gesture was recognized and appreciated by the Lick astronomers.

After his death Keeler was remembered as Adonais, the darling of the gods who was full of promise and talent but struck down before his time. Years later, his colleague William H. Wright wrote about him, "The climax of Keeler's career was probably reached in the Directorship of the Lick Observatory, and the faculties which made for his remarkable success during the short period of his incumbency were both scientific and administrative. . . . It was characteristic of Keeler's method of work that he was always keenly alive to what he was doing." Keeler led by inspiration, Wright said, not by wielding the authority his position commanded. Thus he inspired enthusiasm and loyalty. Had Keeler lived, it is easy to imagine that he would have developed a working spectrograph for the Crossley and obtained spectra of the spiral nebulae, as he wished to do. Whatever he chose to do next, it is certain he would have made more important contributions to astronomy.

8

The Creative Scientist Who Became a

Factory Manager

1900–1923

William Wallace Campbell was director of Lick Observatory longer than any other person, from 1901 until 1930. In his early years as director it was the outstanding astronomical research institution in the world, but in the later years Lick was surpassed by the superior telescope power at Mount Wilson. One of the hardest working astronomers who ever lived, Campbell never stopped working, right to the end of his life.

Campbell was born in 1862 on a farm in Hancock County in northern Ohio, approximately forty miles from Toledo. His ancestors were Scottish, and had arrived in America soon after the Revolutionary War. Wallace, as he was always called, was the sixth and youngest child in his family; his father died when he was only four years old, and he and his three sisters and two brothers were reared by their hard-working mother. They were poor, and Campbell, along with all his brothers and sisters, learned a lot about milking cows and hoeing fields in his childhood.

In school he proved early to have a very good head for figures, and one of his high school teachers encouraged him to enroll in a university, instead of in one of the small colleges that dotted northern Ohio. First, to earn enough money, Campbell had to work for a year. Then in 1881 he entered Ohio State University, but after one semester he was again short of cash, so he dropped out and taught school himself for a few months. In the fall of 1882 Campbell started over at the University of

Michigan as a student in civil engineering. He especially liked working with numbers, and spent much of his spare time expanding his mathematical horizons. In the summer after his junior year he came across Simon Newcomb's book *Popular Astronomy* in the university library. It was so interesting to Campbell that he could hardly put it down; he read it through in two days and two nights. Then and there he decided to become an astronomer. John M. Schaeberle, the professor of astronomy at Ann Arbor, guided Campbell's reading during his senior year and taught him the visual observing and orbit calculation that made up so much of nineteenth-century astronomy.

On graduation from Michigan in 1886, Campbell got a job as professor of mathematics at the University of Colorado. In 1888, when Schaeberle left Michigan to become a member of the initial Lick Observatory staff, Campbell returned to his alma mater and replaced him. This involved a great financial sacrifice, for Colorado was willing to raise Campbell's salary to $2,000 a year in an effort to keep him, while Michigan offered him only an instructorship at $900, but he knew he wanted to be an astronomer and never hesitated.

In 1890, Campbell spent the summer at Lick Observatory as a volunteer assistant and "special student." He received no salary, but was assigned a room in one of the cottages, rent free. In return he had to do one hour's work per day (that is, six hours per week) transferring long columns of numbers from Schaeberle's meridian circle observing books to reduction sheets, a dull, painstaking task. In addition, as practical training, Campbell was allowed to assist James E. Keeler in his spectroscopic observing with the 36-inch refractor two nights per week, to assist Edward S. Holden in his direct photography with it two other nights per week, and to work under Schaeberle's supervision on a photographic program with a small telescope two more nights per week! Campbell thrived on this program. He liked the spectroscopic work best, and by the end of the summer was expert in it. He wanted to come back to Lick the next summer, and Holden was so eager to have him he was even willing to pay Campbell's round trip train fare from Ann Arbor. Before the time came, however, Keeler accepted the directorship at Allegheny Observatory and resigned his Lick position. As soon as he learned that Keeler was thinking of leaving, Holden decided that he wanted Campbell as his replacement. The day that Keeler officially resigned, Holden offered Campbell the job at $1,800 a year, double the salary he was still getting at Michigan, and the young spectroscopist accepted at once.

In June 1891, on his way west to his new job, Campbell stopped off in Boulder. Four years before, while teaching at the University of Colorado, he had been attracted to a sophomore in one of his mathematics classes, Elizabeth Ballard Thompson. She was from Grand Rapids, Michigan, and they had kept in touch and seen each other again on vacations and during the year she had taught at Rockford College, in Illinois, after her graduation from Colorado. Now, his future in astronomy assured, Campbell proposed to her and she accepted. A year and a half later (in the winter, the poor observing season at Mount Hamilton) he returned to Grand Rapids for the wedding ceremony. Throughout the hard-working Campbell's long career Bessie was the major humanizing force in his life. She was younger than he, less scientific and more artistic, less rigid and far more sympathetic to people. She went everywhere with him.

Keeler had done all his spectroscopic research at Lick visually, but photographic plates were rapidly becoming much better radiation detectors than the human eye. Plates can integrate, or store, faint light, and thus in long exposures can reveal much fainter objects than the eye can see. When Keeler went to Allegheny, he designed his new spectrograph for photography and did all his subsequent research with plates. Likewise, at Lick Campbell began working with Keeler's old spectroscope, but replaced its visual eyepiece with a camera lens and plateholder.

With this instrument, Campbell began a series of pioneering investigations of the spectra of hot stars, nebulae and the "new star" Nova Aurigae, which quickly made his name famous to astronomers. He had little formal education in astrophysics, but he was a demon of energy, astronomical spectroscopy was a brand new field, and with the largest refracting telescope in the world and the clear, steady Mount Hamilton sky, he made one important discovery after another. In publishing his results, Campbell always stated the facts as he saw them, with little concern for diplomacy. If a distinguished European astronomer, working with a small telescope under often cloudy skies, had previously published poor data on one of the objects Campbell was studying, the eager young American showed no hesitation in correcting his errors, often with brutal frankness. The aggravating thing was that he was so often right, the elder high priests soon learned.

Holden rightly considered the measurement of stellar "velocities in the line of sight," or radial velocities, as we say today, the most important program of the observatory. Campbell determined them by measuring the "shift" (increase or decrease) of all the wavelengths in a star's spectrum, which gives directly the radial velocity. Photography made it possible to get a permanent record of the spectrum of the star, a "spectrogram" on which this "Doppler shift" could be accurately measured with a precision laboratory instrument. Hermann C. Vogel had begun this type of work with a small telescope at Potsdam in Germany, but Campbell had all the technical advantages at Lick.

He designed a new spectrograph for the radial-velocity program. It was specialized, much less flexible in use than Keeler's earlier instrument, but far superior for the single purpose of measuring Doppler shifts. It was extremely rigid, insulated and temperature-controlled, so that it would not expand or contract from early evening to dawn. These improvements made it possible to take long exposures of faint stars without blurring their spectrograms. Lick Observatory received only a very limited appropriation from the state—enough to pay salaries but not to provide for expensive new instruments—and there were no national foundations that supported research in astronomy. However, Holden persuaded Darius O. Mills, the wealthy financier who had been a member of the first Lick Board of Trust, to give the observatory the money to build the new spectrograph.

With the Mills spectrograph, Campbell and a corps of assistants he trained began a systematic radial-velocity program. It was long, tedious work, but the results were important. By measuring the velocities of many stars all over the sky, Campbell could determine the average motion of our sun with respect to the stellar system. Repeated accurate radial velocity measurements also revealed many "spectroscopic binaries," which appear in the telescope as single stars but are actually pairs in orbital motion about one another. Although too close to be resolved by even the largest telescope, their spectra often showed two sets of lines, with shifts varying regularly, indicating the stars' motions about one another. Even if one member of the pair was much brighter than its partner, so that only one spectrum could be seen, its periodically varying Doppler shift revealed the orbital motion about the unseen companion star.

Success in such a program demands not only observational skill and a large telescope on a favorable site, but also persistence, organizational ability, and unflagging energy. Campbell possessed all these in abundance. He worked single-mindedly at measuring radial velocities during the remaining years of Holden's directorship, and for the all too short two years under Keeler. The program was a great success.

Thus when Keeler died unexpectedly on August 12, 1900, University of California President Benjamin Ide Wheeler immediately placed Campbell in charge of Lick Observatory as acting director. He was the obvious choice as the permanent successor. Campbell stated that he would accept the post if it were offered to him, but that he would not seek it. He sent Simon Newcomb, George Ellery Hale and others a list of the qualifications an ideal director should have—which sounded remarkably like his own—and waited. Wheeler wanted no repetition of the open contest between factions on the Board of Regents that had occurred two years before in naming the next director. He polled twelve of the outstanding astronomers of the world, including Newcomb and Hale, for their advice. All twelve unanimously recommended Campbell. Wheeler proposed his name, and his only, to the regents. They appointed him to the post at their December meeting, and on January 1, 1901, Campbell became the permanent director of Lick Observatory. He was thirty-eight years old, and he was to continue in the office for almost thirty years.

Newcomb, who needed data on the motions of the stars for his analyses of the solar system and of the universe, was especially enthusiastic about Campbell's radial velocity measurements. At the dedication of Yerkes Observatory in 1897, the older man had presented a paper on his new method of determining the average distance of a group of stars by comparing their "proper motions" (apparent angular velocities) with their radial velocities. Today it is known to astronomers as the method of statistical parallaxes. Campbell and Lick Observatory were providing exactly the observational results that Newcomb wanted and needed.

Toward the end of 1900, Newcomb was asked as a foreign associate of the Swedish Academy of Sciences, to recommend a candidate for the new prize in physics that had just been established by the will of Alfred Nobel. The "grim dean of American astronomers," as Walter S. Adams remembered him, decided to nominate Campbell. There was no Nobel Prize in astronomy, but Newcomb considered the young Lick spectroscopist's research as close enough to physics to qualify. He urged Campbell to get out a paper quickly, summarizing all his radial velocity results, that could be included in his list of publications. Newcomb considered Campbell's earlier astrophysical papers on nebulae, Wolf-Rayet stars (with broad emission bands in their spectra), and Nova Aurigae as valuable back-up material but not sufficiently important to earn him the award. Campbell did not get the first Nobel Prize in physics; that award went to Wilhelm Roentgen for his discovery of X-rays. Nevertheless Newcomb's nomination shows how fundamentally important he considered Campbell's radial velocity work.

The new director realized that for a complete statistical discussion he needed radial velocities for stars over all the sky, including the far southern regions that could not be observed from Mount Hamilton. He therefore planned a southern observatory to be erected temporarily in Chile, whose climate is very similar to California's. Instead of a refractor like the 36-inch, Campbell decided to build a reflector. Keeler's work with the Crossley had convinced him and the other Lick astronomers that reflecting telescopes were the ideal research instruments for the future, as David Gill had predicted in his report to the Lick trustees years before.

Campbell drew up a plan to build the telescope and spectrograph, transport them to the Southern Hemisphere, erect them there, and operate them for two years. He calculated the costs carefully (one of his chief assistants later complained that as director of Lick Observatory Campbell exhibited all the skills of a good grocery store manager). According to his estimates, the needed funds would total $24,000. He added 10 percent for unforeseen contingencies, and wrote a letter to Mills, who had moved to New York and established himself on Wall Street. In his letter Campbell explained the research needs of Lick Observatory and how the planned telescope would meet them. Although Mills had provided the much smaller amount needed for the spectrograph several years before, he had subsequently turned down requests for larger sums from Holden and from Keeler. However, Campbell somehow succeeded where his predecessors had failed. Mills ignored the contingency fund and said that if Campbell could accomplish his plan in two years for $24,000, he would give the money. Campbell accepted with alacrity and the "Mills Expedition" was on.

Campbell designed the Mills telescope, as it was called, as a single-purpose Cassegrain (two-mirror) system, to be used with the spectrograph permanently mounted at the focus just behind the primary mirror. The optics were made by the John A. Brashear Company, which for many years supplied instruments for practically all American research astronomers. Brashear had made both Keeler's spectrograph and the Mills spectrograph for Lick Observatory. However, he encountered difficulties in producing the mirrors for the new Mills telescope. It was the first large Cassegrain telescope his firm had attempted. The original plan was to use a spare primary mirror from the Crossley reflector, but it broke while the central hole, necessary for the Cassegrain arrangement, was being bored through it. Then when Brashear completed the new mirrors, Campbell tested them and found them to be unsatisfactory. He rejected them and sent them back to Brashear for further figuring. Campbell spent most of the autumn of 1902 at Brashear's shop in Allegheny, Pennsylvania, so that he could urge on the workmen himself, and personally test the mirrors on stars. In spite of all his drive, cloudy weather prevented the tests and thus the mirrors were not completed in 1902. Campbell returned to Mount Hamilton in December with the promise that James B. McDowell, Brashear's right-hand man, would bring the mirrors to clearer Southern California in January and finish the work on them there.

Campbell had sketched out the general idea of the mounting for the telescope, and Harron, Pickard & McCore, a San Francisco engineering firm, had provided the detailed design. The mounting was built by the Fulton Engine Works in Los Angeles, assembled at Mount Hamilton, and shipped to San Diego, a site Campbell had picked for the final figuring and testing of the mirrors because it is the clearest region of California in winter. There, on January 12, 1903, while Campbell was helping assemble the telescope for testing, he suffered a very serious accident. A heavy piece of the steel mounting slipped, tilted forward, fell on him and bent him double like a jackknife. It nearly crushed him. He was severely injured and his back was permanently damaged. He spent nearly three weeks in a hospital in San Diego, almost completely helpless, before he could return to Mount Hamilton, lying prone on a berth in the train. He had to stop one night in Los Angeles to rest, and again in San Jose, to recuperate for several days before going on to his home. Campbell was in pain for months; it would come back to haunt him when he was tired for the rest of his life. But McDowell finished the mirrors satisfactorily this time, as Campbell's assistant William H. Wright confirmed when he tested them in the telescope in early February in San Diego.

Wright shipped the optics and mounting north and reassembled and aligned the telescope at Mount Hamilton. Campbell had hoped to take it to Chile himself, but as a result of the accident he was still only barely mobile and could not travel. Instead he sent Wright and Harold K. Palmer, who as a graduate student had assisted Keeler at the Crossley. Palmer completed his Ph.D. requirements the day before they sailed. Their ship left San Francisco February 28, 1903, and reached Valparaiso (which had been James Lick's home 70 years before) in mid-April. The telescope, its spectrograph, and a ready-made dome, disassembled and boxed for the trip, came on the ship with them. Wright and Palmer spent one month investigating possible observatory sites before selecting Cerro San Cristobal, a low hill little more than a mile from the center of Santiago. Wright and Palmer had the telescope, spectrograph, and dome erected and assembled before the end of September, and by early October they had taken spectrograms of twenty stars. This was less than three years from the date Campbell had written Mills, asking for the funds for the southern telescope. The total cost of the expedition was less than a twentieth of the cost of the 36-inch refractor and the observatory on Mount Hamilton. Yet the cheap, quickly built Mills reflector collected every bit as much light as the "largest, most powerful telescope in the world" on Mount Hamilton.

The purpose of the Lick southern telescope was exactly the same as that presently served by the Cerro Tololo Interamerican Observatory, erected as a United States national observatory in Chile sixty years after the Mills expedition. Each was built to supplement a northern hemisphere telescope and thus to provide complete coverage of the entire sky.

As the observational results flowed in from Cerro San Cristobal and Mount Hamilton, Campbell was able to use them to calculate accurately the motion of the sun with respect to the stars around it. In other words, he was determining how the sun moves with respect to the "local star system," the part of the galaxy in which it lies. He reported on the differing average velocities of the different types of stars. His team of observers found from their observations that a large proportion of apparently single stars are actually binaries in orbits about one another. He became one of the most respected astronomers in the world, and established Lick Observatory's reputation for careful, exhaustive, accurate observational work. It was a hard-facts factory, and Campbell was its manager.

As observatory director, Campbell also ruled over Mount Hamilton, the little company town where all the astronomers and staff members lived with their families in university-owned houses and dormitories. The main transportation up and down the mountain was the daily stage. The observatory children attended the one-room school, and at recess played "Run Sheep Run" and "Kick the Can" with unlimited territory and wonderful hiding places. Sometimes visitors from town called the energetic children "mountain goats." For the adults, Campbell and his wife presided benevolently over socials, graduations, and picnics. Sometimes the adults presented plays in the school house, or organized dances in the halls of the Main Building. The astronomers laid out a nine-hole golf course on a fairly level area just below the top of the mountain, and Campbell usually finished high in the yearly tournament, though he seldom won it. In summer, some of the families would escape the heat of the exposed mountaintop by going down Mount Hamilton a few miles to "Camp Poodie." There they could enjoy the outdoor life for a few days in a beautiful grove of trees, living in tents, cooking over open wood fires, and letting their children swim in a wide "hole" on Ysabel Creek.

California is earthquake country, and the Hollister fault line runs through Mount Hamilton. Each year many small earthquakes occurred there, and sometimes larger ones too. The famous San Francisco earthquake of April 18, 1906, was relatively mild at Lick Observatory. It occurred early in the morning, and many of the mountain residents felt it as it awakened them in their beds. Most of them ran out of their houses, but the only damage that occurred was that some bricks fell from some of the chimneys, and a few kerosene lamps tipped over and were broken. But the children, walking off to school that morning saw an enormous cloud of black smoke rising over San Francisco. Years later one of them said that it had looked almost like an atomic bomb cloud. The telephone line to Mount Hamilton was knocked out by the earthquake. The daily stage did not arrive from San Jose at noon that day. There was no direct word from the valley, but the observatory people realized that there must be a big fire in the city. Campbell was in Washington at a meeting of the National Academy of Sciences, but that evening the other astronomers turned the 12-inch telescope down below the horizon and through it could see the miles of fire raging across San Francisco. The luxurious Lick House was completely destroyed by the earthquake and fire, never to be rebuilt.

The most severe earthquake in the history of Lick Observatory occurred on July 1, 1911. There was no preshock or audible sound in advance of the quake. It was a Saturday afternoon, and several of the graduate students were lounging in the library when it struck. One of them, Carl K. Kiess, was actually reading the book *The Physics of Earthquake Phenomena* when he felt the floor move and saw piles of other volumes come spilling out of the library shelves. The quake lasted about fifteen seconds, as many of the astronomers, who were used to counting time, reported. The pendulum of a large precision clock in the hall of the observatory shook loose and smashed through its glass case. Many brick chimneys were shaken down. The 36-inch refractor moved about three-quarters of an inch to the south on its foundation, but within forty-eight hours the observatory workmen had moved it back into its correct position with jacks, and it was in operation again. But the old Brick House suffered severe damage. Most of its windows shattered, and nearly all the plaster in the interior of the rooms was shaken loose and fell down. Luckily no one was seriously injured; most of the residents had run outside or dived under tables. Some of the walls of the Brick House were left bulged by as much as four inches.

The building appeared obviously unsafe in another earthquake, or even in a high wind. Everyone moved out of the Brick House immediately, some living in tents for the rest of the summer. University building inspectors came up to Mount Hamilton and confirmed that the old house could not be rebuilt. It was demolished and in its place a much more practical, reinforced concrete dormitory was erected. At the time of the 1911 earthquake Campbell was again away from Mount Hamilton, traveling in Europe with his family, and the assistant director who reported the damage to him wrote that perhaps he better not leave the mountain again, if this was what happened when he went away.

In addition to the radial velocity work, one of Campbell's continuing research interests was his spectroscopic study of the atmosphere of Mars. When he was a young astronomer at Lick Observatory at the end of the nineteenth century, it was generally accepted that William

Huggins, the English astronomer, had detected small amounts of water vapor in the Martian spectrum. The red planet with its reported "canals" was a subject of great popular interest, fanned by Percival Lowell and his writings of a dying Martian civilization. Campbell himself had first observed the spectrum of Mars at the favorable opposition of 1894, when the planet was close to the Earth and well placed for study. The observation is a very difficult one, for the light from the planet goes through the Earth's atmosphere, where there is a considerable amount of water vapor. Its effects must be eliminated, or at least minimized, by careful comparison measurements of another object that does not have any atmosphere, namely the moon.

Campbell's result was that there was no detectable water vapor in the atmosphere of Mars, contrary to what Huggins had reported. Campbell was working with a much larger telescope, the 36-inch refractor, on a high, dry site, Mount Hamilton, with much less water vapor above him in the Earth's atmosphere than Huggins had above his observatory in London. Campbell was sure he was right and said so, loudly and clearly. He was soon embroiled in controversies with Huggins, with Vogel, and with the American spectroscopist Lewis E. Jewell about the amount of water vapor on Mars. The young Lick observer continued to insist there was very little, if any, and he held to his views against all their objections.

After he became director, Campbell returned to this problem in 1908, when Mars again was favorably located for observation. He realized that, to eliminate as completely as possible the disturbing effects of the Earth's water vapor, the ideal observing site would be completely above the atmosphere. This was impossible in the days before artificial satellites, but Campbell did the next best thing and organized an expedition to 14,495-foot-high Mount Whitney in southern California, the highest peak in the continental United States. In August it is one of the driest sites in the world.

Campbell and his assistant Sebastian Albrecht took with them a 16-inch telescope, a spectrograph, and a coelostat, a mirror system that could be arranged to follow the motion of Mars in the sky and bring its light into the apparatus. With them went the Lick Observatory carpenter, to put the equipment together on the summit; a doctor, to preserve their health; a Weather Bureau expert, to measure the humidity; and a local guide to get them to the top. The whole party traveled by carriage from Lone Pine, the town nearest Mount Whitney, to the foot of the trail. There they mounted horses and rode up to their base camp at 10,300 feet elevation, packing in the telescope, spectrograph, and coelostat with them on mules. At their base they camped for three nights to acclimatize themselves to the altitude, and then pushed on to the summit. In two nights there they obtained a series of spectrograms that, when measured later at Mount Hamilton, conclusively showed there was very little if any water vapor in Mars' atmosphere, much less than Huggins or Vogel had reported.

These results, like all the work Campbell did, stood the test of time. In later years even lower limits to the amount of water vapor were set with larger telescopes and better spectrographs, and today space probes have confirmed that there are only minute traces of water vapor in the Martian atmosphere.

As director, Campbell courted the regents and other wealthy friends of the university, whose support he needed for the observatory. He kept after Mills, politely but insistently, by letter and in person every time he went east. In 1903 stock values were down and millionaires felt poor, but in 1907 stocks were up again and Campbell got the money for a new, improved Mills spectrograph. It incorporated many new mechanical features, conceived by Campbell and Wright on the basis of their experience with the original Mills spectrograph. The new one was even more rigid than the previous version, and the spectrograms obtained with it were correspondingly even better. It became one of the most productive instruments in the history of Lick Observatory.

Another of Campbell's special friends was Phoebe Apperson Hearst, the immensely rich widow of millionaire Senator George Randolph Hearst. The first and for many years the only female regent of the University of California, she showered gifts on professors and programs that she particularly liked. Elizabeth Campbell wrote Hearst often and begged her to visit Mount Hamilton whenever she could, while Wallace constantly kept her informed of Lick Observatory's triumphs and needs. She responded handsomely, making many gifts to the university for specific astronomical purposes. In addition she frequently sent fruits, vegetables, and grapes from her estate to the Campbells, as well as Christmas gifts for their sons each year. On one memorable occasion in 1908 she gave Campbell a car. It was the first automobile that he, or anyone else on the mountain, had ever owned. It made travelling up and down the mountain easier for him, and cemented his sons' friendships with all the other Mount Hamilton boys to whom the director often gave rides.

In 1913 Mary McDonald married astronomer Ralph E. Wilson in the first wedding ceremony held on Mount Hamilton. The bride's father, John McDonald, had come to the mountain in the days of the Lick Trust to help build the observatory. He had stayed on for twenty-two years as foreman of the work crew, until his death in 1910. Mary McDonald had been the first child born of parents living at Mount Hamilton; her mother brought her up the mountain from the hospital

Joseph Henry

Laurentine Hamilton

George Davidson

Simon Newcomb

Main Building and Astronomer's House 1888.

Richard S. Floyd and the 36-inch lens 1887.

James E. Keeler

Edward S. Holden

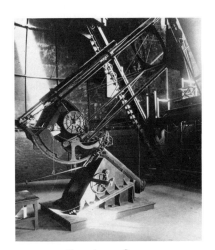

Crossley reflector

S. W. Burnham

Lick Observatory group 1886—Richard S. Floyd standing at left; Thomas Fraser and James E. Keeler (in derby) to left and right of Cora L. Floyd; Edward S. Holden seated with hat to right; Harry Floyd at far right.

Mount Hamilton Picnic, about 1890. John M. Schaeberle lying in middle, James E. Keeler (in hat) with back to camera, Augustus Burnham (also in hat) to his right, Armin O. Leuschner (in dark coat) toward left, Ned Holden (a child in soft hat) to right of him. The three girls are S. W. Burnham's daughters.

Grand dining room of the Lick House, with tables removed, showing the parquetry floor James Lick designed, and the paintings of California scenes, which he framed.

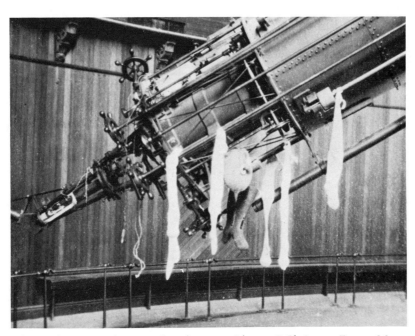

"The stockings were hung on the telescope with care." Christmas Eve on Mount Hamilton, 1927.

John M. Schaeberle and 36-inch lens on the telescope.

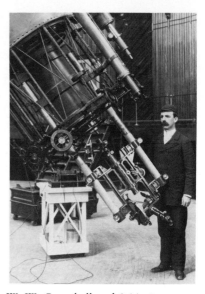

W. W. Campbell and 36-inch spectroscope on the telescope.

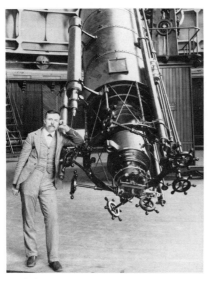

E. E. Barnard and 36-inch telescope eyepiece and controls.

Armin O. Leuschner, Lick Observatory's first graduate student.

36-inch telescope

Lick Observatory group 1899. Left to right, Harold K. Palmer, W. C. Pauli (janitor), Ernest Coddington (behind Pauli), William J. Hussey, R. Tracy Crawford, W. H. Wright.

Lick Observatory group at door of Main Building 1900. In rear, left to right, Cora M. Keeler, W. H. Wright, James E. Keeler, Ernest Coddington, R. Tracy Crawford, Frank E. Ross, William J. Hussey, Robert G. Aitken. The umbrellas are held by (left to right) W. W. Campbell, Charles D. Perrine, and Harold K. Palmer. In front, Harry Floyd is seated to left of the two Keeler children.

Visitors and horse drawn stages outside the Main Building about 1900.

W. W. Campbell at the wheel of the automobile Phoebe Apperson Hearst gave him, outside the Main Building. In the back seat are bride Mary McDonald and groom Ralph E. Wilson, just after the first wedding on Mount Hamilton, 1913.

Lick Observatory eclipse group at Jeur, India, 1898. W. W. and Elizabeth B. Campbell (in sun helmets) at far left, adjusting a spectrograph. The Schaeberle camera, in the background, points to the sun.

Lick Observatory Main Building in the 1950s.

Lick Observatory group in 1922, just before Campbell assumed the presidency. He is in center, Joseph H. Moore at far left, and W. H. Wright and Robert G. Aitken standing third and second from right, respectively.

W. W. Campbell, president emeritus of the University of California and director emeritus of Lick Observatory, 1933.

Lick Observatory group in 1960. Seated in front, starting third from left, Constance Watson, Albert E. Whitford, C. Donald Shane, William P. Bidelman, George H. Herbig. Seated in second row from left, Joel Stebbins, Nicholas U. Mayall, and Stanislavs Vasilevskis. In rear, Merle F. Walker is sixth from left, Gerald E. Kron, fourth from right.

James E. Keeler at time-service switch-
board 1886.

Joel Stebbins at 36-inch spectrograph
1903.

W. H. Wright seated at desk in Lick
Observatory director's office 1941.

C. Donald Shane explaining model of
120-inch reflector to Governor Earl
Warren 1946.

Hole left by crash of airplane into Main Building, 1939. A few of the fragments of the plane had not yet been removed.

Left to right, Donald E. Osterbrock, Albert E. Whitford, Robert P. Kraft.

120-inch mirror after first aluminization, with Nicholas U. Mayall, Dwight J. Ludden, Stanislavs Vasilevskis, and Howard Cowan, rear to front.

120-inch Shane reflector

120-inch Shane reflector dome

in San Jose when she was only three weeks old. Campbell gave the bride away in the ceremony, held in the library of the Main Building, and afterward personally chauffeured the wedding party to San Jose.

A few years later the horse-drawn stage that made daily round trips from San Jose to Mount Hamilton was replaced by an automobile. It was an early, primitive model, with a cooling system unsuited to the long climb up the mountain. On hot summer days, one passenger recalled years later, the stage driver would have to refill the radiator frequently along the twenty-six mile route, not only at the regular stops at Grand View, Hall's Valley, and Smith Creek, but also on emergency "pauses" at little springs along the steepest parts of the road. By 1914 Wright had the second privately owned automobile on Mount Hamilton, and a few years later a third appeared.

During World War I Campbell was a superpatriot. He agonized over President Woodrow Wilson's "lack of backbone" in not getting America into the war earlier. Once the United States was in the fight, Campbell gloried in boasting about how many of the observatory's sons were serving in the armed forces, and how fully everyone on Mount Hamilton supported the Liberty Loan bond drives. All three of the Campbell sons served in the war, Wallace (known in childhood within the family as "Mowgli," a name from a Rudyard Kipling story) in a medical unit in Europe, Douglas as a pursuit pilot who became one of the first American aces, and Kenneth, the youngest, as a Navy recruit who had just begun training when the Armistice was signed. Campbell had quickly come to hate the Germans during the war, and remained unreconciled to them for years. Long after most other American scientists had accepted their former enemies back as fellow research workers, Campbell fulminated in his letters about the imperialistic "Teuton" scientists who in his opinion were "awaiting the opportunity for revenge, and will be ready to encourage their country to go to war again as soon as they think they can win." Although many English, Swedish, French, and Swiss astronomers visited and worked at Lick Observatory after World War I, no Germans came while Campbell was director.

A hundred-percent American and one of Campbell's favorites on the Lick Observatory faculty was Charles D. Perrine. He had come to Mount Hamilton as the observatory's secretary in Holden's time. Perrine's previous experience had been in a business office in Alameda, but he was interested in astronomy and photography, and eager to get a job at Lick. He volunteered for observing duties in addition to his secretarial work, and was soon taking daily direct exposures of the sun for a routine survey of sunspots. He helped with the nighttime observing on the 36-inch, and working on his own with one of the smaller telescopes, he discovered a comet. He longed to become a full-time astronomer. In response, in 1895 Holden changed Perrine's title to secretary and assistant astronomer, but left his salary at the same $1200 a year. Keeler raised Perrine's pay to $1500 in 1898, but kept him at both jobs. Finally, after Keeler's death, Campbell got another secretary and put Perrine in charge of the Crossley reflector. Many of the pictures of nebulae and galaxies in the Keeler memorial volume, published in 1908, were actually later, better photographs that had been obtained by Perrine.

He also took nearly a thousand direct exposures of the asteroid Eros during its close approach to the earth in late 1900 and early 1901. This was part of a cooperative program to measure directly the distance to this minor planet, which has a highly elliptical orbit that brings it very near the earth at rare intervals. The measurement of this distance provides a direct determination of the astronomical unit, the scale of the solar system.

Perrine supervised a drastic reconstruction of the rickety Crossley reflector during the years between 1902 and 1905. He had the mechanical mounting, drive, and gears all improved greatly. He removed the traditional Newtonian flat mirror, which brought the light out to a focus at the side of the tube. Instead he mounted the plateholder directly at the prime focus of the telescope, in the middle of the upper end of the tube. Perrine introduced a system of prisms and transfer lenses so that the observer could "guide," or accurately follow the motion of the stars during the exposure, from an eyepiece just outside the tube. Thus with only a single reflection in the light path to the photographic plate, the Crossley became an even faster and more efficient instrument for obtaining pictures of nebulae and star fields.

With the improved Crossley reflector Perrine discovered the sixth moon of Jupiter in 1904, and the seventh in 1905. Both are small, faint, and distant from the planet. He recognized them as satellites of Jupiter by detecting their motion with the planet through the background of neighboring stars on exposures he took on successive nights. The regents responded to these discoveries by promoting Perrine to "astronomer," the equivalent of professor, and by raising his salary $200 a year. Santa Clara University awarded him an honorary Sc.D. degree. In 1907 Perrine left Lick Observatory to become the director of the Cordoba Observatory in Argentina. He was one of the last of the old-time astronomers, who, like Burnham and Barnard, was able to achieve important research results on the basis of sheer drive, technical skill, and interest with only a minimal scientific education.

Campbell believed in results, and another of his favorite astronomers was Heber D. Curtis, surely one of the most productive and creative scientists who ever worked at Mount Hamilton. Like many of the early

Lick astronomers, Curtis was a product of the University of Michigan, but he had been a student of Greek, Latin, Hebrew, Sanskrit, and Assyrian in the department of classics while Schaeberle and then Campbell were teaching astronomy across the campus. After earning a master's degree at Ann Arbor in 1893, Curtis got a job as professor of Latin and Greek at Napa College, a small Methodist institution north of San Francisco. When it ceased operations he moved on to the University of the Pacific, located near San Jose and close to Mount Hamilton. The University of the Pacific had a small Clark refracting telescope, and Curtis began using it; soon he was teaching astronomy and mathematics instead of Latin and Greek. He came to Lick as a special student during his summer vacations in 1897 and 1898, and in the summer of 1899 returned to Michigan to learn orbit computations.

Although he was married and had a child, Curtis decided to quit his job, make a mid-career switch, return to graduate school, and become a professional astronomer. Keeler would not accept him as a regular student at Lick, but Curtis got a fellowship at the University of Virginia where he earned his Ph.D. after only two more years of study. Then Campbell hired him, immediately after his graduation in 1902, as an assistant on the Lick staff. At first Curtis worked on the radial-velocity program; in 1906 Campbell sent him to Chile to replace Wright as head of the southern station. In both posts Curtis developed and improved his skills with astronomical instruments. Like Keeler, he was an expert observer with a creative imagination and a questioning mind, who continually identified and attacked new problems.

When Curtis returned from Chile in 1910, Campbell put him in charge of the Crossley reflector. He began studying the nebulae, as Keeler had before him, taking better long exposures, as well as photographic spectra. One group of objects, the planetary nebulae, were easily identified by their symmetric forms and their emission-line spectra, which indicate that they consist of hot glowing gas. Their nature had first been recognized by the English observer Huggins years before Lick Observatory had even been founded. Because many of these "planetary nebulae" appear roughly circular, they could be mistaken for planets when seen through a small telescope. That and their greenish color, similar to Uranus', is the origin of their name. Curtis took many photographs of known planetary nebulae with the Crossley reflector, compiling sequences of pictures that showed their structure at various levels of brightness. He integrated these images into carefully prepared master drawings of each object, which he published together with a description of their spectra. Many planetary nebulae appear to be rings in the sky, and Curtis, by photometric measurements of his plates, showed that in fact they are toroidal figures, not spherical shells seen in projection as some astronomers had postulated. Curtis' monograph on the planetary nebulae, published in 1918, remained for decades a landmark paper in our knowledge of these objects.

Even more important, however, was Curtis' research on the spiral "nebulae," objects quite unlike the planetaries. He confirmed, as Perrine had earlier, Keeler's result that long-exposure photographs show many of these spirals of all sizes, from a few large objects down to almost countless small ones just barely large enough to be resolved. As Curtis wrote in 1918:

> "It is my belief that all the many thousands of nebulae not definitely to be classed as diffuse or planetary are true spirals, and that the very minute spiral nebulae appear as textureless disks or ovals solely because of their small size. Were the Great Nebula in Andromeda situated five hundred times as far away as at present, it would appear as a structureless oval about 0.2 long, with very bright center, and not to be distinguished from the thousands of very small, round or oval nebulae found wherever the spirals are found. There is an unbroken progression from such minute objects up to the Great Nebula in Andromeda itself; I see no reason to believe that these very small nebulae are of a different type from their larger neighbors."

Keeler had estimated the number of spiral "nebulae" within reach of the Crossley reflector as 120,000; Perrine, with better plates, had raised this estimate to 500,000; now Curtis, with still better working material, made it 1,000,000.

Furthermore, in studying the forms of the spirals as revealed on his direct photographs, Curtis recognized that they are circular, flat objects, more or less disks in form. The few seen nearly edge-on appeared dark in their central planes; he could only attribute this to "occulting matter" (today we say "dust") in these zones. He suggested that these occulting effects were the result of the same general cause that produces similar effects in our galaxy, such as the "coal sacks" and "dark nebulae" discovered by E. E. Barnard in the Milky Way. When Curtis published this study in 1918, he did not quite come out and say that the spiral "nebulae" were not really nebulae, that is clouds of gas presumably condensing into stars as the old cosmologists had suggested, but instead were really incredibly distant galaxies, massive star systems like our own. However, his letters show that he had believed it for years; it was the only sensible interpretation of his data.

In 1919 Curtis published a very full statement of his picture of spiral nebulae as galaxies or "island universes" composed of "hundreds of millions" of stars. One very strong point in favor of this idea, Curtis wrote, is that in the sky the spirals are all distributed outside the Milky Way (toward its poles); he could find none actually "within" it (near the galaxy's central plane) even on very long exposures with the Crossley reflector. Curtis' natural interpretation was that the "occulting matter" in the plane of the Milky Way, which prevents us from seeing distant stars, also prevents us from seeing any spiral nebulae in these directions. Therefore they must all be distant objects. The radial velocities that had been measured for a few spirals were much higher than the velocities of stars, indicating clearly that they belonged to a different class of objects, Curtis said. Furthermore, if the spirals were really as close to us as stars, their high velocities would carry them across our line of sight at a rate that would lead to easily observable "proper motions," but Curtis had compared his Crossley photographs of many spirals with Keeler's photographs taken years before, and no such motions were detectable. Still another proof, in Curtis' view, was provided by the occasional novae or "new stars" detected in some spirals. Most of them were much fainter than the novae known in our galaxy, indicating that they (and therefore the spirals in which they were located) were far outside our galaxy. Finally, Curtis noted that the spectra of the spirals do not show bright lines characteristic of a hot gas, like the planetary nebulae. Instead they have continuous spectra with absorption lines, as do stars and star clusters, known to be composed of "vast congeries of stars."

In 1920 Curtis presented his views in the form of a lecture before the National Academy of Sciences in Washington, as part of what came to be known as "the Great Debate." His "opponent" (it was not really a debate, but rather two talks given in succession by two different speakers) was Harlow Shapley of Mount Wilson Observatory. He had been studying the globular clusters in our galaxy, and had used them to measure its size. The value Shapley found for its diameter (3,000,000 light years) was so large that he believed the spiral "nebulae" must belong to the galaxy. Curtis, however, emphasized the analogies of the spirals to our galaxy, their spectra, and the novae observed in them as evidence that they are separate, distant star systems. Both Shapley's and Curtis' data were poor by modern standards. Shapley's distances were too large because he did not know of interstellar extinction, and some of Curtis' distances were wrong because he did not recognize the difference between novae and supernovae. Nevertheless, it is clear today that Curtis was basically right, and Shapley was basically wrong.

That is hindsight; at the time the correct conclusion was not so obvious. Curtis was sure he was right, but he did not convince Shapley. Many other astronomers simply followed the conservative path and continued to believe what they had learned as students, that spirals were "nebulae," and thus within our galaxy. A few years later Edwin Hubble, with the 100-inch telescope at Mount Wilson, was able to identify Cepheid variables, which are clearly individual stars, within spirals and thus to prove Curtis' point.

Just a few months after "the Great Debate," Curtis left Lick to become director of Allegheny Observatory in Pittsburgh, a more highly paid position with more responsibility. However the frequently cloudy weather, bright sky, and poorly equipped telescope prevented him from accomplishing much research there. As he wrote Campbell, "the California combination of instruments PLUS climate is a hard one to beat." He confessed that "there is no place like the hill [Mount Hamilton] for astronomical work and that any man who leaves these opportunities is bound to be sorry for it," just as Campbell had predicted he would be when he left.

Ten years later, Curtis moved on to the University of Michigan where he had begun as a classics student decades before. He was amused by its 37½-inch reflecting telescope, which had been built by William J. Hussey, his predecessor at Ann Arbor and the man who in 1897 had led the rebellion over the Crossley reflector at Lick Observatory against Edward S. Holden. "As you know H[ussey] was not a practical man" Curtis wrote Campbell. "This [Michigan] reflector is surely unique in all the world, and has more wild ideas and inconveniences than any I know. I consider it worse than the original Crossley mounting."

Curtis hoped some day to raise the funds to build a large new reflecting telescope for the University of Michigan, first projected as an 84-inch, later a 96-inch. However, he believed it would be "a scientific crime" to locate it in the Midwestern climate, and hoped to reach an arrangement by which it could be erected on Mount Hamilton. The Great Depression made it impossible to raise the necessary funds until long after Curtis' death in 1942, but a generation later the University of Michigan did build first a 60-inch, and then a 100-inch, both on Kitt Peak in Arizona.

Curtis had justified the glowing words that Campbell, never given to hyperbole, had written about him in 1918: "Dr. Curtis is one of the finest men personally, and one of the finest scientifically, whom I have been privileged to know. He is unusually adaptable to conditions existing around him, unusually quick in sizing up the situation, whether personal or scientific; he has excellent mechanical training and ideas; he is a good mathematician, physicist and photographer, and he is very resourceful in the devising or use of scientific apparatus for solving new problems."

In his own long career as director of Lick Observatory, Campbell became one of the most respected and influential scientists in America. He was awarded honorary doctor's degrees by seven universities, beginning with the Western University of Pennsylvania (now the University of Pittsburgh) in 1900 and ending with the University of Chicago in 1931. He received many decorations and medals, including the Henry Draper Medal of the National Academy of Sciences and the Catherine W. Bruce Medal of the Astronomical Society of the Pacific. He served at various times as president of the Astronomical Society of the Pacific, the American Astronomical Society, and the International Astronomical Union.

Campbell was always a forceful, effective leader. In 1922 when the regents of the University of California began looking for a new president to take over the institution the following year, they naturally thought of Campbell. He had had no academic experience outside of Lick Observatory, and was initially reluctant to accept the post. However, the regents persuaded him, and in 1923 he moved to Berkeley as president. Before taking the position, he made it a condition that he would retain the directorship of the observatory as well. He would set the general policy, and Robert G. Aitken would be in charge of day-to-day operations on Mount Hamilton as associate director. Campbell knew that the regents desperately wanted him as president, but he wanted to be sure he had a job to go back to if he failed in Berkeley.

In fact Campbell soon learned how to handle undergraduates, football players, and department chairmen on the campus almost as well as he had handled night assistants, astronomers, and influential scientists on the mountain. He remained president of the University of California until he retired in 1930. Then he was elected president of the National Academy of Sciences. He and his wife moved to Washington, and he took over his new office in the summer of 1931. By then the Great Depression was in full force. Campbell was nearly seventy years old, very conservative, and increasingly out of sympathy with the national government, particularly after Franklin D. Roosevelt became president in March 1933. Campbell battled with the Democratic administration for what he regarded as the prerogative of the Academy, to be the chief source of scientific advice for the nation. He was not successful, and by 1935, when his term ended and he and his wife returned to California, he had been worn down by the fight.

Campbell wanted to return to Mount Hamilton, but he realized that he was too old to start his research career again, and that there was no place for him there. He and his wife lived on for three years in San Francisco. His old friends from Lick visited him on their trips to the city, and discussed astronomy with him. One of the last things he did was to read Princeton astrophysicist Henry Norris Russell's new book on *The Solar System and Its Origin*. Campbell annotated his copy of it in detail, criticizing Russell for giving too much attention to eastern and English astronomers, and not enough credit to the Lick staff, especially to Keeler. In particular, Campbell believed that Russell had gotten the story of the water vapor content of Mars' atmosphere all wrong, systematically downgrading his own contributions.

By then Campbell was blind in one eye, losing the sight of the other, and suffering frequently from aphasia. He feared that he would become a burden for his wife and his friends. On June 14, 1938, he committed suicide, ending one of the most productive lives in the history of American science.

9

In the Shadow of the Moon

1889–1930

W. W. Campbell was a great ob-
servational astronomer. The spectroscopic radial velocity program that
he planned, organized, and directed was the backbone of Lick Observatory research for almost thirty years. In addition, throughout his long
life he carried out another program that took him all over the world,
far from Mount Hamilton. It was the solar eclipse program, which also
made Lick Observatory famous in the astronomical world.

Total solar eclipses are rare events that occur under very special circumstances. They can take place only when the moon passes directly
across the face of the sun, rather than just above or just below it in the
sky as it usually does. By coincidence, the size of the moon as seen from
the earth is almost exactly the same as the size of the sun, so when
they are perfectly centered at a solar eclipse, the moon completely blocks
out the bright light of the sun for a few brief moments. The coincidence
is so close that the eclipse is only total along a narrow path, which may
lie anywhere on the earth, swept out by the line from the center of the
sun through the center of the moon. Such an eclipse darkens the sun
and provides the only chance astronomers have to study its outer layers,
the chromosphere and the faint corona, and the regions of space close
around the sun that ordinarily are hidden in the dazzlingly bright sunlight scattered by the earth's atmosphere.

Today, trained solar astronomers fly halfway around the earth, taking
their highly specialized eclipse instruments with them, to collect data at
these rare events. In the early days of Lick Observatory astronomers
were far less specialized than they are today, and almost every researcher felt himself qualified to observe an eclipse. So little was known

about the sun and its immediate surroundings that almost any new observational fact was likely to be valuable.

The first Lick Observatory eclipse expedition was organized by the first director, Edward S. Holden, for an eclipse very close to home. It was the total solar eclipse of January 1, 1889, whose track of totality cut across northern California. To Holden there was never any doubt that Lick Observatory should send a party to observe it. He himself, as an astronomer at the United States Naval Observatory, had led one of the many groups of astronomers who observed the solar eclipse of July 29, 1878. Its path of totality cut across the Rocky Mountains. Holden and his party, which included young James E. Keeler, then a student between his first and second years at Johns Hopkins University, had observed the eclipse from Central City, a mining town high in the Rockies above Denver.

A few years later Holden, by then director of Washburn Observatory at the University of Wisconsin, had headed a party that observed the total solar eclipse of May 6, 1883 from Caroline Island, a tiny atoll in the Pacific Ocean, midway between Hawaii and Tahiti. For that eclipse he had journeyed over 12,000 miles by commercial steamer, naval vessel, and train, stopping off to visit Mount Hamilton on his way back to Madison. At both these eclipses, during the few moments of darkness, Holden had searched the sky with a small visual telescope to see if he could pick up any possible planet with an orbit even closer to the sun than Mercury's. The existence of such an intramercurial planet had been predicted on theoretical grounds by Urbain Leverrier in 1859, and there was even a name waiting for it, Vulcan. James C. Watson had thought he had spotted it, and another interior planet as well, at the 1878 eclipse, which he had observed from Separation, Wyoming, but the scientific world was skeptical. In 1883 Holden found no intramercurial planet, and concluded that none as bright as magnitude 5.5 could exist, or he would have seen it.

He was especially eager to send a Lick expedition to the January 1, 1889 eclipse because astronomers were coming across the country to California from Harvard and from Washington University in St. Louis to observe it; surely the local scientists could go a few miles to do the same. Holden, always anxious to build support for astronomy and for Lick Observatory, wrote and distributed a pamphlet of suggestions on eclipse photography for amateurs. Thirty teams, organized by the Pacific Coast Amateur Photographic Association, set up their cameras at Cloverdale, on the track of totality in Sonoma County, and photographed the eclipse according to these instructions. Holden used the interest created by the eclipse to start the Astronomical Society of the Pacific on February 7, 1889. For many years it was dominated by Lick Observa-

tory, but it continued to flourish and grow. Today it is the largest combined amateur-professional astronomical society in the United States.

For the site of the Lick Observatory eclipse camp Holden picked Bartlett Springs, in the Coast Range a few miles north of Clear Lake. He
did not go himself, but assigned Keeler to head the Lick group. The
other members of the party were E. E. Barnard, Charles B. Hill, the
observatory secretary, and Armin O. Leuschner, Lick Observatory's first
graduate student. Keeler, Barnard, and Leuschner left Mount Hamilton
for the site on December 15, two weeks in advance of the eclipse. Ned
Holden, the director's thirteen-year-old son, was to accompany them
as a volunteer observer. They traveled from San Jose to San Francisco
by train, but Ned managed to get lost on the platform in San Jose.
Leuschner stayed behind, found him, and brought him on to the city
on the next train. They remained there two days, assembling supplies.
They succeeded in persuading Ned to join a group of amateurs, headed
by a friend of Holden's, who were going to watch the eclipse from
Cloverdale, and thus got the director's son off their hands. The Lick
astronomers then went on by narrow-gauge railroad to Sites, the nearest
station to Bartlett Springs. There they discovered that their astronomical
instruments, which they had boxed and sent ahead over a week previously, had not arrived. Keeler went to Colusa, a larger town with an
express office, to try to find them, while Barnard and Leuschner continued on to Bartlett Springs. It had been raining for days and the whole
area was a sea of mud. They traveled by a conveyance that was called
a "stage," but was in fact an open wagon. The road was uphill all the
way to Bartlett Springs, and the all-day trip left them tired, hungry, and
soaked. Keeler found the boxes, sent them to Bartlett Springs, and came
on himself the next day in the same open wagon.

Bartlett Springs was a little mountain resort, with one hotel and a
group of small cottages. Barnard picked out a clear area in the middle
of the croquet court for their telescopes and cameras. He and Leuschner
dug holes in the mud for the piers while waiting for Keeler to arrive
with the instruments. It rained more or less continuously until December
28, when the skies cleared long enough for them to set up and adjust
their telescopes. Keeler, assisted by Hill, was to observe the spectrum
of the corona visually with a 6-inch Clark refractor. Barnard was to
operate a battery of cameras, the largest built around a 3-inch aperture,
49-inch focal-length lens that had been borrowed from a telescope ordinarily used to check the water level in a distant reservoir at Mount
Hamilton. Leuschner was to measure the brightness of the corona directly, comparing its light with that of a standard candle.

They all rehearsed their observing procedures every time the rain
stopped, until they were letter perfect in their assigned tasks. The night

before the eclipse the skies clouded over, and it looked as if there would be more rain or even snow the next day. However, on January 1 the sky was clear, and they were able to carry out all their observations. Keeler was struck by the very great brightness of the continuous spectrum of the corona. From the variation he observed in its brightness as the eclipse progressed, he concluded that this continuous spectrum actually was emitted in a region around the sun, but that the emission lines were more probably emitted in the chromosphere, close to the sun, and then were diffracted (bent) by the edge of the moon. This was a theory that had been put forward by Charles S. Hastings, his former physics professor at Johns Hopkins. Keeler was right about the continuous spectrum, but wrong about the emission lines, as he realized when he compared his measurements with the reports of other observers. A year later, in summarizing the results of the Lick eclipse observations, Holden gave the correct interpretation, that both the continuous spectrum and the emission lines were really emitted in the corona, and that Hastings' theory was therefore disproved.

Barnard's direct photographs came out very well. The exposure he had taken with the long-focus lens showed a wealth of fine detail in the corona. To the Lick astronomers it seemed much the best picture taken at the eclipse, superior even to the one obtained a few miles away at Willows by William H. Pickering, the Harvard astronomer, with a larger and much more expensive telescope. Barnard's picture demonstrated that photography was far superior to the older method of drawing to record the physical appearance of the corona at an eclipse. From then on, every Lick eclipse expedition featured a long focal-length camera to photograph the chromosphere and corona.

The next Lick Observatory eclipse party consisted of S. W. Burnham and John M. Schaeberle, the other two of the original four faculty members. The eclipse was on December 22, 1889. Its track of totality crossed Africa and South America, and Holden decided to send the expedition to Cayenne, in French Guiana. There were no funds in the Lick budget for travel, and although this had not been a problem for nearby Bartlett Springs, it was a serious one for Cayenne. However, Charles F. Crocker, a prominent banker and member of the Board of Regents, provided the necessary money as his contribution to science.

In addition to two smaller cameras, Schaeberle took an 18-inch parabolic mirror that he had made himself to photograph the corona. Its focal length was 12½ feet, providing an image of the sun 1.3 inches in diameter, considerably larger than those produced by any of the cameras Barnard had used at the earlier eclipse. At Cayenne, Schaeberle built a wooden tube for this reflecting telescope from barrel hoops and boards. He made it so that the telescope could be slowly turned sideways an inch and a half to follow the motion of the sun during the two minutes of total eclipse. Burnham and Schaeberle, both conservative astronomers of the old school, made no spectroscopic observations.

This expedition was far from successful. Cayenne, in the tropics, was very humid, and the Lick astronomers had a constant battle to keep their equipment dry. They had not allowed enough time to build, assemble, and test the telescope before the eclipse. The skies were cloudy most of the time, and as a result Schaeberle was not able to focus the reflector in advance, but had to do it visually the very day of eclipse. Burnham, a dedicated amateur photographer who believed in freedom of expression, tried to estimate the exposure times as the eclipse occurred rather than following a predetermined program based on the measured characteristics of the photographic plates. As a result all the pictures were badly overexposed. Furthermore, the roughly constructed reflector could not be moved smoothly, and this together with the imperfect focus contributed to the poor definition of the resulting images of the corona.

Nevertheless the large scale of the photographs allowed much detail to be seen. The pattern of coronal filaments, or streamers, so familiar to us today was clearly revealed. Schaeberle developed a ballistic theory of the corona on the basis of these photographs. According to his ideas the streamers were the trajectories of material ejected from sunspots. He claimed his theory showed that it was not necessary to assume electric and magnetic forces to understand the corona. In this he was quite wrong, for he did not realize that magnetic lines of force, coming up in and near sunspots, channel the charged solar plasma, but his idea that trajectories of matter are involved had some validity.

The published report on this eclipse, especially the parts written by Burnham, contained open criticism of the plans that Holden had drawn up for the expedition. Burnham was particularly biting in his comments on "the so called Carcel standard lamp" with which the director wanted them to calibrate their plates for photometric measurements. Holden himself, in his introduction to the report, replied to his staff members' criticism. This was the first public manifestation of the struggle which led to Burnham's bitter resignation and departure from Mount Hamilton in 1892.

In Chile, at Mina Bronces on Cerro Cobre in 1893, Schaeberle first used the 40-foot focal-length camera, which became a standard instrument that was taken on all Lick eclipse expeditions from then until 1932. Its lens, 5 inches in diameter, had been made by Alvan Clark and Sons for a heliometer, a fixed horizontal telescope that was used with a flat mirror for taking daily photographs of the sun at Mount Hamilton. This lens provided an image of the sun (or of the edge of the moon, at an eclipse) just over 4 inches in diameter. It would show the fine coronal detail in even larger scale than the reflector Schaeberle had used at Cayenne.

For the Chile eclipse, which would last only a few minutes, his plan was to mount the lens rigidly on a concrete pier, and the photographic plate at the focal plane on another pier forty feet behind it. The telescope thus could not move. It had to be erected so that it was pointed accurately to just where the sun would be at the moment of eclipse, but calculating this alignment was no problem for a trained astronomer of that era. During the three minutes of totality the sun's image would move six inches in the focal plane, so the photographic plate had to be driven by clockwork along a fixed track at just the right rate to compensate for this motion. The telescope had no rigid tube, but black canvas was strung between the objective lens and the plateholder to keep out stray light. This canvas was held by a wooden framework, anchored to the ground but completely free of the concrete piers.

Schaeberle assembled this fixed telescope and tested it at Mount Hamilton in the autumn of 1892. The altitude of the sun at the eclipse would be 24 degrees, so a slope with that approximate gradient was needed to mount it. This would be no problem in Chile, nor was it at Mount Hamilton.

Holden managed to raise the money to support this eclipse expedition from Phoebe Apperson Hearst. In spite of her great fortune, she believed in rigid economy in her gifts, and the funds she provided were barely sufficient to send one person all the way to Chile and back. Schaeberle went alone, beginning the Lick tradition of counting on local people recruited at the scene to help with the instruments. He left San Francisco by steamship on January 25, 1893. Three months later he was at Mina Bronces, a copper-mine settlement at an elevation of 6600 feet, in the Atacama desert about 200 miles north of Santiago. The dry climate and thin mountain air would reduce the scattering of sunlight in the earth's atmosphere, and thus enhance the contrast of the coronal features.

Schaeberle's volunteer assistants included several mining officials, three British naval officers from warships anchored at nearby Carrizal Bajo, and three local amateur astronomers. They operated the four smaller cameras Schaeberle had brought with him. For days in advance of the big event he drilled them all in their assigned tasks. April 16, the date of the eclipse, was beautifully clear. Schaeberle himself worked with the 40-foot camera, changing the plates rapidly according to a prearranged plan, so that he could take a total of eight exposures, ranging from ¼ second to 32 seconds. One of the assistants counted seconds, beating them out with a stick on an empty box, during the 2 minutes and 53 seconds of totality. Schaeberle and his volunteer helpers took all their scheduled exposures. He spent ten days developing them (there were fifty plates in all, and the eight he had taken with the 40-foot camera were each 18 × 22 inches in size) and making glass positive copies of all of them.

These large-scale plates showed the most detailed structure recorded in the corona up to that time. To Schaeberle they seemed to confirm his ballistic theory of the coronal streamers. In addition he thought that he had discovered a comet very close to the sun on two of his plates. Actually it may have been a large and unusual condensation within the corona.

The next eclipse to which a Lick Observatory expedition went was on August 9, 1896. W. W. Campbell wanted very much to go to that one, and Holden had promised him that he could. However, toward the end of 1895, the young spectroscopist had to inform his director that he could not make the trip. His wife had become pregnant, in spite of their old-fashioned "planning," and he did not want to be far from home when the baby came. Schaeberle went instead. The site chosen was Akashi, Japan, on the southwestern part of the island of Honshu, not far from Kyoto and Osaka. However, it turned out lucky for Campbell that he had not gone, for the skies were completely cloudy in Akashi on the day of totality. Schaeberle's long trip across the Pacific was in vain. Two months before the eclipse the Campbells' second son, Douglas, was born; twenty-two years later he was to become one of America's first World War I aviation aces.

Campbell got to observe his first eclipse at Jeur, India, in 1898. He and his wife had dreamt of sailing around the world; this eclipse gave them their chance to do it. They left Mount Hamilton on October 16, 1897, just after Holden had given up the Lick directorship and departed for the East. The Campbells traveled westward by ship via Honolulu, Tokyo, Kobe, Nagasaki, Shanghai, Singapore, and Ceylon, arriving at Bombay nearly two months after they had cleared the Golden Gate. Holden and Campbell had decided on a site at Karad, but on arrival in India the Lick astronomer learned that an epidemic of bubonic plague was raging there. Instead he went northeast to Jeur, on the flat, dry plain of Deccan. He and the Mamlatdar of Karmala, the local chief government official, scoured the countryside in a two-wheel pony cart to select the best site.

Campbell was an extremely capable astronomer, and an extremely self-confident man. But at Jeur, worn down by travel, heat, tension, strange food, and strange customs in a strange land, he found that he could not determine the exact latitude and longitude of the site he had chosen. His observations and calculations did not check to the last decimal point. He suffered something like a temporary breakdown, ranting at his difficulties and doubting that he would succeed in observing the eclipse. To his wife he seemed "wild-eyed" and "talking as I had never heard him talk before." She prescribed a massive dose of scotch and soda (she had brought the ingredients with their supplies) and put him to bed. Though he tossed and turned all night, by morning he was cured. He found their location on a map and saw it was near an English survey station. This put them within two miles of the center of the fifty-mile wide path of totality. Campbell knew he would have no difficulty observing the eclipse, and he realized that the time signals on which he had been depending, received over an Indian telegraph line, must have been in error. His confidence returned, apparently never to desert him for another forty years.

One of Campbell's reasons for choosing Karad as the observing site originally had been that it was in a hilly region, where the Schaeberle camera (as it was now called) could be set up on a convenient hill. Jeur was in a flat desert, but the 40-foot tube had to be erected at an elevation of 51 degrees. Campbell solved this difficulty by having native workmen dig a pit down to bed rock, ten feet deep in the ground. The plateholder and its mechanism were installed at the bottom of the hole. The lens was mounted 24 feet above the ground on a triangular-shaped wooden tower that Campbell built himself, because he could find no local carpenter who measured up to his own exacting standards. Around this tower he had the workmen pile up a high earthen wall, topped by a second, larger, completely independent wooden tower. Canvas screens were tied around the outer tower, to keep the inner one, which supported the lens, from being buffeted by the wind during the exposures. The tube of the telescope, which served only as shield against stray sunlight, was also supported independently of the lens and plateholder.

Campbell trained a force of volunteers to man the instruments at the eclipse. Captain Henry L. Fleet, a short, stout, John Bull-like character who commanded the Royal Marines at Bombay Harbor, operated the Schaeberle camera, along with Engineer T. Garwood of the Royal Navy. Major Samuel Comfort, the U.S. consul at Bombay, was in charge of the short-focal length telescope used for photographing the outer corona; Rev. J. E. Abbot, an American missionary who had studied astronomy under Charles A. Young at Dartmouth years before, worked the shutter of a special six-prism high-dispersion spectrograph that Campbell had assembled to measure the wavelength of the strong green line in the spectrum of the corona. He and Elizabeth Campbell handled two other spectrographs themselves, and other volunteers helped by calling out the time. Campbell drilled the entire group for four days before the eclipse. On January 22 the sky was clear, as it had been for six weeks, and although Campbell had a bad headache he was not nervous at all. All the observers carried out their duties flawlessly, except for one lieutenant who forgot to uncap a spectrograph. Campbell consoled him, but no doubt resolved to drill everyone a little longer before the next eclipse.

The daytime temperature was in the nineties, far too hot for developing the plates. Campbell waited until after midnight, when the outside temperature had dropped to 42°F, before beginning the photographic work in his temporary darkroom, one tent pitched inside another. It took a week for him to develop all the spectrograms and direct plates, working from 1 A.M. to sunrise each morning. He found that all the instruments had worked satisfactorily, and that the observational program had been a complete success. Then they quickly packed and struck their camp, moving out in a long line of bullock carts.

From Jeur the Campbells continued their trip around the world. Before leaving India they saw Jaipur, Delhi, Agra, and the Himalayas. They continued by ship from Bombay to Suez, then overland to Cairo. On the mainland of Europe they visited observatories in Rome, Florence, Milan, Nice, and Paris, and in England Campbell went to Greenwich, Tulse Hill, Kensington, Oxford, and Cambridge before they sailed for home from Liverpool on May 4, 1898. War had been declared between the United States and Spain, but they were traveling on an English ship and were safe on the high seas.

After their arrival in New York, the Campbells went on to the midwest and visited his relatives in northern Ohio, and her family in Grand Rapids, Michigan. Then he returned to California, while Elizabeth stayed behind in Iowa for two weeks more, visiting other relatives. Campbell got back to Lick Observatory seven and a half months after he had departed on the eclipse expedition. He reached San Jose just in time to ride up Mount Hamilton in the stage on June 1, 1898 with the new director, James E. Keeler, on the very day he took over the observatory.

Campbell did not find time to begin measuring and analyzing the eclipse plates until winter. Summer and fall were the good observing seasons, and he devoted all the time he could to observing at the telescope. When he measured the high-dispersion spectrogram of the corona in January, he found that the wavelength of the strongest emission line was 5303.26 Å. Until then this line had been assumed by most astronomers to be due to magnesium, which has a strong line that is seen in absorption in the spectrum of the sun at wavelength 5317 Å. Most of the strong absorption lines of the solar spectrum appear as emission lines in the chromosphere, which is just inside the corona and like it is observed at eclipses, so it was a natural mistake to make. Campbell's measurement showed that this assumption was wrong, and gave the correct wavelength. Unfortunately for him the English astrophysicist J. Norman Lockyer had also obtained a high-dispersion spectrum of the corona at the same eclipse. He had not waited to measure his spectrogram, so he, not Campbell, first discovered this incorrect identification and published the right wavelength.

Comparing the wavelengths of the green line in the corona on the two sides of the sun, Campbell was able to detect a very slight difference. Clearly it resulted from the rotation of the sun. The velocity of rotation he measured from this Doppler shift, approximately 3 km/sec, showed that the corona has essentially the same period of rotation as the main body of the sun, about 25 days.

The next easily accessible solar eclipse was to be on May 28, 1900. Its path of totality ran through the southeastern United States, and practically every American observatory and astronomy department planned to send observers there. Samuel P. Langley, the pioneer solar astrophysicist who had become Secretary of the Smithsonian Institution, had been greatly impressed with the direct photographs of the corona that Campbell had taken with the Schaeberle camera in India. Langley planned to use an even longer focal-length telescope in 1900, and he wanted Campbell to join his group and operate it. Keeler, however, succeeded in raising money from William H. Crocker to finance a Lick expedition, and Campbell headed it.

With him he took Charles D. Perrine, the Lick Observatory secretary who had become a research astronomer. Heber D. Curtis, who was going east to start his graduate-student career at the University of Virginia, joined them as a volunteer. They located at Thomaston, Georgia, a tiny hamlet that Weather Bureau statistics showed was in one of the clearest areas on the track of totality. But the food was very poor in the little Southern town, and Campbell and Perrine suffered constantly from diarrhea. Perrine was so ill that even the hard-driving, perfectionist Campbell insisted that he remain in bed for several days before the eclipse. Luckily Curtis proved very handy with the instruments, and Campbell depended heavily upon him. No doubt Curtis earned his future job at Lick by his outstanding performance at this eclipse.

In addition to the Schaeberle camera, again set up with its objective lens on a tower and its plateholder in a pit as at Jeur, they operated several spectrographs. Two of them were "moving-plate" instruments of a type first used by Campbell at the 1898 eclipse, and further improved for this one. The idea was that as the shadow of the moon moves across the sun, just at totality, it successively covers and uncovers layers at different heights in the chromosphere. Thus a time-resolved picture of the spectrum can furnish information on just where in the chromosphere each spectral line is emitted. To provide this time resolution the photographic plate was driven by a clockwork mechanism, so that it moved slowly and smoothly upward in the focal plane of the spectrograph, widening the spectrum into a broad band. Each level in this band corresponded to a particular instant of time during the eclipse, and thus showed the spectrum of the range of heights in the chromosphere uncovered at that moment.

Several volunteers assisted Campbell, Perrine, and Curtis at the eclipse, including two Dutch astronomers, the mayor of Thomaston, and the president of the Robert E. Lee Institute, a small private military school. Campbell drilled them all for three days before the eclipse, until they became completely proficient in changing plates and opening and closing shutters at exactly the preassigned times.

On May 28 the Weather Bureau proved correct, but only just barely. The sky clouded over completely at sunrise, then cleared, then began to cloud up again, but remained clear over the sun throughout the eclipse. At 7:40 A.M., one minute after the sun had begun to peek out from behind the edge of the moon, the sun was again covered by clouds. The Lick astronomers and their temporary assistants performed their duties flawlessly. One moving-plate spectrograph stuck momentarily, but it was quickly repaired.

It was too hot in their unventilated darkroom to develop their plates that day, but using ice they cooled their developer to 75°F by 10 P.M. Then Campbell, Perrine, and Curtis processed all the plates, working until dawn. Campbell had wanted to get out of Thomaston just as soon as he could after the eclipse, but his departure was delayed by Ormond Stone, the director of McCormick Observatory of the University of Virginia. He was a visual observer of the old school, but he had photographed the solar corona at the eclipse. Now he confessed to Campbell that he did not know how to develop his plates, and asked Campbell to do them for him. "It was the appeal of a drowning man and I couldn't refuse," Campbell reported to Keeler.

Finally, after developing nearly all of Stone's plates and packing their instruments for shipment back to California, Campbell and Perrine managed to get away from their eclipse camp. Writing from Tybee, Georgia, on the Atlantic coast, Campbell informed Keeler:

We left Thomaston yesterday morning with the good will of everybody in the town, and with pleasant recollections, on our part, of everybody save the landlady, who is a terror. Arriving here by the Atlantic, we enjoyed the first square meal we have had since leaving Atlanta on April 31. More than all, the apparent cleanliness of the food here is a pleasant and radical change.

I have been very well, except that I almost constantly paid the penalty of eating unclean and imperfectly cooked food, and as a result am considerably "run down." Mr. Perrine is very much thinner than we have ever seen him at Mt. Hamilton, and still complains of soreness throughout his entire digestive system. But all he needs is something to eat.

The people of Thomaston did for us the best they know how, but they don't know how.

Although nearly all their photographic plates came out well, Campbell described in detail in this letter the three moving-plate spectrograph exposures that had been spoiled. Keeler, who knew Campbell was going to an astronomical meeting and would see many scientists on the way home, cautioned him:

> While I agree with you that important results should not be concealed, I would advise dwelling on the successes rather than on the failures. If you were to tell a reporter that three plates out of the ten were failures, he would receive a totally different impression from what he would if you gave him the equivalent statement that seven out of ten plates were successes.
>
> I am glad to learn that you found something to eat at last. It must have been in the "Cracker" country. I have been there myself, but never long enough to starve.

After a stop at Charlottesville to develop Stone's last two plates, Campbell and Perrine proceeded to New York for the meeting of the Astronomical and Astrophysical Society of America (now the American Astronomical Society) that had been scheduled for the month after the eclipse. There they learned just how successful they had been. Langley's Smithsonian party had gotten their coronal exposures with a 135-foot focal-length horizontal telescope, and they showed just about as much detail in the corona as did the Lick plates taken with the Schaeberle camera. Edwin B. Frost and E. E. Barnard of Yerkes Observatory had also obtained good spectra and direct exposures at their Wadesburo, North Carolina eclipse site. But most of the other astronomers and physicists who had descended on the eclipse track had not planned thoroughly enough, had not allowed enough time to adjust their instruments carefully, and had not rehearsed their mechanical tasks for days before the eclipse. As a result most of them had failed to get usable data. Robert W. Wood, then a very promising young experimental physicist, who observed the eclipse with the Naval Observatory party, called the event "the Pinehurst fluke . . . it was awful."

Another eclipse took place a year later in May 1901. This time the path of totality ran through Sumatra, in the Dutch East Indies. It was too far for Campbell, who had just become director of Lick Observatory, to go himself, but he sent Perrine in his place. At Padang, the site of the Lick station, the ground was too wet and soft to dig a firm pit for the plateholder of the Schaeberle camera. Hence for the first time the entire telescope was mounted above ground, and as the sun was near the zenith, the double tower, made of local bamboo, was particularly high. Totality lasted for the unusually long time of six minutes and nine seconds, perfect for long-exposure photography. In addition to the long-focus camera for recording the corona at large scale, Perrine used a special wide-field camera to search photographically for a possible planet within the orbit of Mercury, as Holden had done visually in 1878 and 1883. This instrument was called the Vulcan camera, the name Leverrier had given the hypothetical planet years before.

At the time of the eclipse thin cirrus clouds covered the sky, but Perrine managed to obtain good photographs through them. The longest exposure with the Schaeberle camera showed the coronal features well. No intramercurial planet was detected.

In 1905 Lick Observatory sent out not one but three eclipse parties. The eclipse of August 30 had an usually long track, and Campbell's plan was to have observers take photographs in Labrador, in Spain where the eclipse would occur an hour and a half later, and in Egypt, where it would happen still another hour later. Comparison of the long-focus images from the different sites would show motions in the corona, if they occurred, while similar comparison of the wide-field photographs would reveal the motion of any intramercurial planet, allowing its orbit to be calculated immediately. Campbell, in charge overall, went to Spain with Perrine, while Curtis, accompanied by Joel Stebbins, went to Labrador, and William J. Hussey traveled to an island in the Nile near Aswan. Elizabeth Campbell went with her husband, with the responsibility of supervising the living and eating arrangements, as she had in India and as she was to do on all subsequent eclipses that he observed. He never again complained about the food.

Unfortunately, however, on the day of the eclipse, Curtis and Stebbins were completely clouded out in Canada, Campbell and Perrine had to take their photographs and spectra through thin clouds in Spain, and Hussey observed under conditions of poor seeing (the temperature was 108°F) and a very bright sky (the result of fine sand in the air) in Egypt. The plates were not very good, but set some limits to the speed of coronal motions and to the brightness of any intramercurial planet.

Two and a half years later the next Lick eclipse expedition, headed by Campbell, had better luck at Flint Island in the central Pacific Ocean. The large group, which included the Campbells, Perrine, Robert G. Aitken, Sebastian Albrecht, and physicist E. P. Lewis of the Berkeley campus, sailed by commercial steamship to Tahiti. From there the U.S.S. *Annapolis*, a navy gunboat, carried them five hundred miles northwest to a little atoll that was one of only two landfalls along the entire track of totality. Again, as in Sumatra, the Schaeberle camera was mounted above ground, nearly vertically, in the midst of a group of smaller cameras and spectrographs. The Lick astronomers practiced their assigned tasks for nearly a month before the date of the eclipse, January 3, 1908. That morning the skies alternately cleared and clouded over several times. Less than a minute before the beginning of totality, rain was falling, but it stopped just in time. The clouds quickly dissipated, and during the last half of the nearly four minutes of total eclipse the sky was completely clear. The seeing was excellent, and the Lick observers secured an excellent series of direct exposures and spectrograms.

When they were developed the plates showed no object as bright as eighth magnitude within 25° of the sun, corresponding to no Vulcan larger than 30 miles in diameter. To explain the motion of the perihelion of Mercury as a gravitational perturbation, the basis of Leverrier's prediction, would have required more than a million such planets, and they obviously could not all have been missed. Perrine declared the search for an intramercurial planet ended.

To make up for dropping this search, after 1912 Campbell started another far more important eclipse program. It was nothing less than an astronomical test of Albert Einstein's theory of relativity. Einstein had put forward his special theory in 1905, and in 1911 he published a paper working out its predictions concerning the effects of gravitation on light. Einstein showed that a massive body (like the sun) bends light rays passing near it, much as it deflects the paths of particles. The predicted bending or deflection decreases outward from the body. Its maximum value for stars whose rays just pass the limb of the sun, according to the special theory of relativity, is $0''.87$, a very minute angle. Planets and indeed all other objects have similar effects, but since the amount of deflection is proportional to the mass of the body, it is predicted to be too small to measure, even for Jupiter, the largest planet. The only chance to observe the deflection of light rays by the sun is to measure the positions of stars very near it at an eclipse, when the sky is dark, and to compare them with their positions at another time in the year, when the sun is gone from the field. Einstein himself suggested this measurement in his paper.

The German astronomer Erwin Findlay-Freundlich took up the problem. His first idea was to measure existing plates, taken at previous eclipses, to see if this effect was present. He naturally wrote Campbell, as well as astronomers at other observatories, to ask for glass copies of their eclipse plates. This was in 1912, before World War I, and Campbell was glad to oblige. He put Curtis to work on the problem. They agreed that the intramercurial planet search plates, taken with the Vulcan cameras, would provide the best chance of success, since they showed many stars over a wide range of angular distances from the sun. Curtis selected several of the best plates from the Spain and Flint Island eclipses, copied them, and sent them to Findlay-Freundlich. He measured them, but found that the errors of measurement were so much larger than the predicted effect that it was hopeless to try to test the theory with the existing plates. Clearly better data were needed. Campbell decided that Lick Observatory should provide them.

He tried to do so at the next suitable eclipse, on August 21, 1914. Its track of totality was in Russia, and the site Campbell chose was in the little town of Brovary, twelve miles northwest of Kiev. Campbell made this eclipse expedition a family trip with a vengeance, taking along his wife, his mother-in-law, his three sons (then students at Harvard and at Hotchkiss), and a Harvard friend of theirs. In addition Campbell took Curtis to help with the observations. They rented a dacha, or country house, where once again Elizabeth Campbell presided over the domestic arrangements.

The four Vulcan cameras were arranged on a single mounting. Each one was to photograph a field 5°-by-6½°, centered on the sun, on a 16 × 20 inch plate. World War I had begun, and Russia started mobilizing its army after the party reached Brovary, but the American astronomers were allowed to continue their preparations for the eclipse. On August 2, Germany declared war on Russia, and the members of the nearby German eclipse expedition were all arrested. The older Germans were deported, but Findlay-Freundlich and the younger men were held as prisoners of war. (They were later exchanged for an equal number of Russians who had been caught in Germany.) Campbell, Curtis, and the family were prepared to photograph the eclipse on August 21, but the skies were completely covered by thick gray clouds and no observations were possible. The Lick group returned safely but empty-handed. It was the first time Campbell had been completely clouded out at an eclipse, and he was keenly disappointed. He and the others managed to get out of Russia via Moscow, Finland, Sweden, and Norway, going around Germany on their way back via London to America, but they could not take their eclipse cameras and spectrographs with them, and had to leave them with Russian astronomers for safekeeping.

In 1916 Einstein announced his new general theory of relativity, in which the equivalence principle (between accelerations and gravitational fields) was a fully developed keystone. It predicted a gravitational deflection of light of the same form as the special theory, but exactly twice as large. Hence the necessity for an observational test was obvious.

Campbell, to his great credit and unlike many of his scientific contemporaries, approached the general theory of relativity with an open mind. He was too old and too out of touch with theoretical physics to understand it in any depth, but nevertheless he felt that it should be tested, rather than denounced, or declared invalid on the ground that it seemed contrary to "common sense." This is just what quite a few experimental physicists and observational astronomers were doing at that very time.

Campbell hoped to measure the gravitational deflection of light at the eclipse of June 8, 1918, at Goldendale, Washington. All but one of the lenses for the Vulcan cameras were still in Russia, along with the lens for the Schaeberle camera. An American businessman who was a friend of Campbell had located them and the other eclipse instruments at Pulkovo Observatory, where they had been stored for the duration of the war. He started them back to California via Vladivostok and Japan, but there were delays everywhere and they did not arrive until after the eclipse. Campbell borrowed two other lenses from Chabot Observatory in Oakland, and improvised other equipment for the Goldendale eclipse. Instead of the Schaeberle camera, Campbell used a 6-inch, 40-foot focal-length lens for the large-scale direct photographs of the corona.

Curtis, who was teaching navigation at a wartime naval officers' school in Berkeley, got away for a few weeks to adjust and focus the instruments at Goldendale. He was to operate the Vulcan cameras at the eclipse. A large party was present from Lick, as well as two notable figures from its early days, Ambrose Swasey, whose firm had built the mounting for the 36-inch refractor, and John A. Brashear, who had supervised the making of so many spectrographs and lenses for the observatory. Banker William H. Crocker, who had financed several of the eclipse expeditions (including this one) was also there with his wife. Most importantly of all, John Hoover, the foreman of the Lick workmen, had been at Goldendale for over a month before the eclipse, making and assembling all the substitute telescopes, cameras, and spectrographs.

Everyone rehearsed their tasks under Campbell's supervision for two days before the eclipse. The sky clouded over the night before the big event, and remained overcast until one minute before totality, when a rift appeared. A small region, including the sun, remained clear during the entire three minutes of totality, but then clouded over again. The Lick observers had been exceptionally lucky, just at the moment of the eclipse. However, the 6-inch lens produced a "ghost" or double image of the corona that was quite unsatisfactory. Campbell and Joseph H. Moore obtained several good spectrograms of the corona, from which they measured the wavelengths of a total of fifteen emission lines. The plates taken with the Chabot lenses for the light deflection measurements appeared satisfactory, except that the clouds had somewhat weakened the star images farthest from the sun. It was necessary to obtain plates of the same star field at night, without the sun, for comparison with the eclipse plates. This could not be done until several months later. In the meantime Curtis, who was to measure the plates, left California for the Bureau of Standards in Washington, where he was to do technical design work for the duration of the war.

Thus the project languished until Curtis returned to Mount Hamilton in May 1919. By then Campbell had learned that British scientists would send two parties to Brazil and South Africa to attempt to measure the Einstein effect at the May 29, 1919, eclipse. On that date the sun would be located in a rich field of bright stars, the Hyades, ideal for measuring the predicted light deflection. Campbell was anxious for Curtis to complete his measurements so that Lick Observatory could announce its results before the British. Curtis worked hard, and although the star images on the plates were not as small as he had hoped, he managed to complete his measurements before the meeting of the Astronomical Society of the Pacific in Pasadena on June 19, 1919. The result Curtis found was that there was no gravitational deflection of light, and this was announced at the meeting. Campbell, who went to the International Astronomical Union meeting in Brussels that same summer, reported Curtis' measurements but emphasized the possibility of large errors in the result, because of the elongated star images and the relatively small number of stars recorded on the plates. He gave as his own opinion that Curtis' measurements ruled out the larger deflection ($1.''73$ at the limb of the sun) predicted by the general theory of relativity, but not the smaller value ($0.''87$) predicted by the special theory. However, Campbell's discussions with the astronomers he met in England and Belgium led him to realize that the "probable errors" (that is, the range of uncertainty) of the measurements from the Goldendale plates were too large to allow any firm conclusions to be drawn. He cabled Curtis to hold up the paper, and it was never published.

Then on November 6, 1919, at a joint meeting of the Royal Society and Royal Astronomical Society in London, Arthur Eddington announced the result of the British eclipse expedition of that summer. Their measurements appeared to confirm the prediction of Einstein's general theory of relativity. Eddington himself had been at Principe Island, off the coast of Africa. He recorded only a few stars on his plates, from which he found a deflection of $1''61 \pm 0''30$. The other English group, at Sobral, Brazil, obtained more stars on their plates, which when measured gave $1''98 \pm 0''30$. It seemed clear that the general theory was correct. In Britain the idea that their scientists had traveled to the far corners of the Earth soon after a savage war had ended to confirm an abstruse theory proposed by a German especially seized the popular imagination.

Many American astronomers remained skeptical however. The two 1919 results differed between themselves by more than their stated probable errors, and it was only by averaging them that the final value, 1″80, which agreed closely with Einstein's prediction, was obtained. Clearly further confirmation was desirable. Campbell still hoped that it would come from the Goldendale plates. After Curtis left Lick to become director of Allegheny Observatory in April 1920, Campbell handed the Goldendale material over to Robert J. Trumpler, a young Swiss astronomer who had come to America in 1915. He had worked at Allegheny from then until 1919, when he came to Mount Hamilton with a one-year Martin Kellogg Fellowship. When Curtis decided to take the Allegheny directorship, Campbell hired Trumpler on the Lick staff in his place. His training and experience were in positional measurements of stars on photographic plates to determine their "proper motions" (angular motions in the sky), exactly the skills needed for the light-deflection measurements.

At various times in the next year Campbell set Trumpler, Adelaide Hobe and Moore to reanalyzing Curtis' measurements and calculations, or to remeasuring the Goldendale plates themselves. They all found that Curtis had made numerical slips in his haste to get results for the Pasadena meeting, but none of their new measurements or reductions gave much better results. The basic problems were with the Chabot lenses and the second-rate clock-drive systems the Lick observers had been forced to use at Goldendale, because their prime instruments were still in transit from Vladivostock.

Campbell decided to make an all-out attempt to measure the light deflection accurately at the next eclipse, on September 21, 1922. The eclipse would last over five minutes, allowing long exposure times that would record many faint stars. The site Campbell chose, at Wallal on Ninety-mile Beach in northwestern Australia, had a very good record for clear weather. There was no way to get there except by sea, but Campbell arranged that the Australian Navy would bring his party in and land them over the surf in small boats.

He ordered a pair of new lenses from John Brashear & Co., with 5-inch apertures and 15-foot focal lengths, especially designed to give good images over a large field. He also ordered a pair of the newly designed Ross lenses, with shorter focal lengths but even wider fields. Curtis, at Allegheny, tested all the lenses for the Lick group as Brashear's opticians produced them. Campbell sent Trumpler out to the Pacific with the cameras to take check plates of the eclipse star field several months before the event, so there would be no delay afterward. The

Lick director wanted to get his results out first, as he knew that several other parties would also be attempting to measure the light deflection at the same eclipse. Since Wallal was so isolated, off all the normal shipping lanes, Trumpler went to Tahiti in the same latitude as northern Australia to take the check plates. He set up the new Einstein telescopes, as they were called, near the beach at the outskirts of Papeete. Campbell had wanted Trumpler to make all the measurements of these plates in Tahiti before the eclipse, but somehow he never managed to find time to do so, so this part of the program had to wait.

Campbell, Elizabeth, and Moore, who was to handle the spectrographs, sailed from San Francisco, rendezvoused with Trumpler, their telescopes, and an independent Canadian eclipse party at Broome, Australia, and were taken to Wallal by an Australian warship. Once again there were rehearsals, beautifully clear skies at the eclipse, and complete success for the Lick astronomers with their observational program. They developed some of the plates at Wallal and the rest at Broome, their first stop on the way home. Campbell had planned that he and Trumpler would measure the plates there, but a delay enroute left them too little time, and they did not finish the measurements until after they got back to Mount Hamilton.

The care and preparation that had gone into the Wallal expedition paid off handsomely. The plates had good images of many stars, well distributed in distance from the sun. There was great popular interest in Einstein and relativity, and fellow scientists as well as reporters pressed Campbell for an early statement of his results. He held out until he and Trumpler had measured the plates completely, and checked all their calculations carefully. Then he announced the results to the newspapers on April 11, 1923 (and cabled them to Einstein the next day). The new observations completely confirmed the prediction of the general theory of relativity. The measured deflection at the edge of the sun was 1".74. This was the preliminary value based on reductions of only some of the plates, as Campbell gave it in a paper he presented to the National Academy of Sciences later that month. The final value, based on all the plates and published by Campbell and Trumpler in 1928, was 1".75 ± 0".09, very close to the predicted deflection.

This was convincing evidence to nearly all scientists, whatever their nationalities. However it did not convince T. J. J. See, an embittered astronomer at the Mare Island Navy Yard. As a young man he had been a promising theoretician of gravitational astronomy, but he had highly exaggerated notions of his own importance and skills. These brought him into repeated clashes with nearly everyone who tried to work with him. He had been fired from jobs at the University of Chicago and at Lowell Observatory, and then, a few years after securing an appointment at the Naval Observatory, had been exiled to Mare Island.

The day after Campbell released the Wallal results, confirming the relativity prediction, See launched a violent attack on Einstein's theory as "a greater piece of humbuggery [than] has . . . appeared in any age." He was quoted in the *San Francisco Chronicle* as saying that "everybody knows full well that the ether does really exist and act, with forces equivalent to the breaking strength of millions of immense cables of the strongest steel, for holding planets in their orbit." In his written statement See claimed that he valued the work of Lick Observatory, and recognized the importance of the eclipse measurements. However, he said, Campbell was wrong in thinking that they confirmed "the discredited doctrine of relativity." Instead, See stated, the gravitational deflection had been predicted by Isaac Newton and Johann G. von Soldner, who in 1801 had actually calculated the effect under the classical gravitational theory. See quoted Philip Lenard, the notorious anti-Semite who headed the "German physics" movement, to the effect that Einstein had stolen his idea from Soldner but had gotten the numbers wrong when he did so.

All See's facts were wrong. Campbell put Trumpler to work checking out the historical record. In an article in the *Publications of the Astronomical Society of the Pacific* Trumpler convincingly refuted all See's wild charges. Campbell himself issued a strong statement denouncing See's criticisms (without mentioning his name) as "based on prejudice with which I have no sympathy."

By this time Campbell was president of the University of California; the committee of the regents who persuaded him to take the job had met him and first offered it to him as he came ashore from his ship in San Francisco on his return from Australia. In spite of his high office, Campbell fully intended to observe the eclipse of September 10, 1923, whose path of totality cut across Catalina Island (once owned by James Lick) and Baja California, Mexico. He had picked the site, not far from Ensenada, two years before, while he was still a working astronomer. After he became president, Campbell placed his former assistant and long-time associate on the Lick faculty, William H. Wright, in direct charge of the Ensenada expedition. However, as the day of the eclipse drew near, the old warhorse could not keep away. He and Elizabeth drove down to Ensenada with their oldest son, Wallace, where they joined Wright, Moore, Trumpler, Hamilton M. Jeffers (then at the University of Iowa but later a member of the Lick staff), and many others in the large eclipse party. Campbell and Wright had planned an ambitious observing program. However, they saw nothing but clouds and storms for days before and after the eclipse, and had no chance to get any data. Their only consolation was that Catalina, where most of the other eclipse parties had gone, had equally bad weather.

Campbell loved to go on eclipse expeditions. As Wright wrote of him:

> One was never quite sure whether he enjoyed more the intense concentration during the critical moments of totality, or the excitement of preparation and travel incident to the undertaking, which sometimes occupied the larger part of a year. To one living on a relatively isolated mountain top, an opportunity to see the world under extraordinarily interesting circumstances is not a negligible matter, but it may be said of Campbell that, whichever of these two aspects of an eclipse expedition appealed to him the more, he served them both well.

However, Campbell never did use the observational material secured at all the eclipse expeditions to make a serious study of the sun. He regarded it as his duty to obtain the best data he could during the few minutes of total solar eclipse granted each year by nature. He published the main observational results as soon as he could, the measured wavelengths of the chromospheric and coronal emission lines and the direct photographs of the eclipsed sun. These had a real effect on astronomy; indeed the Lick wavelengths were known to be highly accurate and were among the best of those used by European spectroscopists Walter Grotrian and Bengt Edlen when, years after Campbell's death, they identified many of the coronal lines with highly ionized iron and calcium. Likewise the coronal photographs taken with the Schaeberle camera showed the intricate fine structure of the corona that eventually yielded to interpretation as filaments, their forms governed by magnetic forces, in a very high temperature plasma. Campbell probably intended to return to a more thorough analysis of all the eclipse data some day, but he never found time to do so.

The person who did work up Campbell's eclipse spectra was Donald H. Menzel. He came to Lick in 1926 as a young staff member hired specifically to do astrophysical research, and in particular to analyze the eclipse spectra. Aitken, who was by then running the observatory as associate director, had written to Henry Norris Russell, the great Princeton astrophysicist who made so many contributions to all branches of astronomy and spectroscopy, to ask him if he could recommend a good candidate for the job. Lick Observatory needed a person to do astrophysical research, Aitken said, but he could not find one anywhere. Russell recommended Menzel, his former student, whose interests, he said, were "thoroughly astrophysical."

Menzel used Campbell's eclipse plates from 1898 through 1908 for a very thorough study of the chromosphere. This is the name given the region just outside the main body of the sun (which is opaque and appears brilliantly white to the layman, but "yellow" to astronomers), but inside the very faint, tenuous corona (which is only visible at a total eclipse). The chromosphere, although it was first discovered at an eclipse in the nineteenth century, is bright enough to be observed outside of eclipse if the sun's disk is properly occulted (covered). However only its brightest features can be studied in this way, and the best data must be obtained at a total eclipse. The chromosphere has an emission-line spectrum, and its name ("color-sphere") came from the fact that it is very red, the result of the great strength of the hydrogen Hα line. Menzel applied the atomic physics he had learned so well at Princeton to interpret Campbell's plates of the "flash spectrum" at eclipses. The old visual observers used this name because the emission-line spectrum of the chromosphere seemed to them to "flash up" just as the eclipse became total; actually what they were seeing was the bright continuous spectrum of the main body of the sun disappear, leaving the much fainter chromospheric spectrum visible at last. At each eclipse Campbell had used moving-plate spectrographs, to record the flash spectrum as a function of time during the eclipse. Menzel could thus measure the total emission from all the layers of the chromosphere projected above the edge of the moon at each instant. Then by subtraction he could find the emission from each layer alone. This was the starting point for his physical analysis.

One problem was that Campbell had not taken photometric calibration exposures, using a lamp as a light source, its brightness reduced by various amounts in well-measured steps. Such calibration spectrograms are essential for measuring accurately the relative strengths of different emission lines at different levels in the chromosphere from their apparent brightnesses on the photographs. Without the calibration exposures, Menzel had to estimate the strengths of the lines as well as he could. It made his task more difficult, but did not stop him. In Menzel's hands, Campbell's old spectrograms combined with the newest methods of atomic physics yielded important new knowledge about the outer layers of the sun.

Menzel found that the temperature decreases outward from the solar atmosphere into the lower chromosphere, falling 1,000° to a value of about 4,700°K. From that point outward, he found, there are increasingly large deviations from standard thermodynamic equilibrium. The most convincing evidence was the presence of ionized helium in the high chromosphere. The absorption lines of this ion occur only in the spectra of very high-temperature stars. Menzel and everyone else thought that the temperature in the outer chromosphere could not possibly be higher than the 4,700°K of the lower chromosphere, far too cool for ionized helium. The older, traditional astronomers believed that the presence of ionized helium in the outer chromosphere resulted entirely from its low density, but Menzel showed by quantitative calculations that their qualitative reasoning was incorrect. He examined various theoretical ideas as to just how the deviations from thermodynamic equilibrium might occur, showing that none of them agreed with all the observed facts, but that turbulence must somehow be important in the chromosphere.

Menzel completed his long paper on the chromosphere in the summer of 1930. He wished to include in it not only the observational results, but a complete theoretical discussion of the outer solar atmosphere. Aitken, a self-trained double-star observer with no knowledge of the "newfangled astrophysics," counseled Menzel to stick to "the facts" and leave these theoretical speculations out of his long paper reporting on the eclipse data. With the help of his friend J. Robert Oppenheimer, an assistant professor of physics at Berkeley, Menzel had worked out the quantum mechanical equations for the absorption coefficients of atoms and ions, but Aitken insisted that he omit them from the paper. Such theoretical calculations, the associate director felt, did not belong in a Lick Observatory publication. Campbell, who had just retired as president of the University of California, backed Aitken, and Menzel had no option but to accept their decision. He published his paper in the *Lick Observatory Publications* with, as he thought, the heart cut out of it, and he did not like it. Aitken gave Menzel permission to publish the excised part in the *Monthly Notices of the Royal Astronomical Journal*, which he considered a suitable repository for purely theoretical arguments. The paper is, in fact, very good. It demonstrates Menzel's great ability to compare and interweave theoretical reasoning and observational data, getting the most out of both approaches to understanding nature by combining them.

Another unexpected result that Menzel derived from the chromosphere spectra was the high proportion of hydrogen in the sun. It was far more abundant than all the other elements together. Cecilia Payne had found the same result a few years earlier in her work on the spectra of the hot stars at Harvard, but it was contrary to the standard ideas of the time and even the great Russell could not accept it. He talked her into de-emphasizing this result and trying to explain it away. Aitken was perfectly willing to let Menzel publish this conclusion; he had no "common sense" on the subject of the hydrogen abundance of the universe. According to Menzel's later recollection, Russell finally accepted the high hydrogen abundance in the solar atmosphere while visiting Lick and discussing the eclipse results with him.

In his paper Menzel strongly emphasized the importance of turbulence in the outer solar atmosphere. He described the chromosphere as a "hot spot" phenomenon, "the spikes [spicules, in modern terminology] that form it being rooted in superheated areas of the sun's surface." This paper was an important one for the future of solar astrophysics. As Russell wrote Menzel, it "put the whole matter [of the chromosphere] on an entirely new basis." It justified all the observational efforts that had been expended on the Lick eclipse expeditions, even if some of the senior Mount Hamilton staff members were too rigid to accept all its physical conclusions. Menzel himself left Lick in 1932 for a faculty position at Harvard, where he had a long and highly successful career in nebular and solar astrophysics. He was one of the first to realize that the temperature actually *increases* outward in the chromosphere, explaining many of the apparent anomalies he had earlier measured on the Lick eclipse spectra. To the end of his life Menzel always considered himself a protégé of Campbell, the pioneer spectroscopic observer of nebulae and eclipses.

10

Leuschner's Legacy

1888–1931

Graduate students are the lifeblood of research institutions. Young, eager, and dedicated, they are frequently incredibly hard working and are always stimulating to their elders. While they learn, they help to produce research results; as they work, they train themselves to be the independent research scientists of the future.

From its earliest days Lick Observatory was a center of graduate work in astronomy, and over its long history it produced many of the outstanding research astronomers in America. Of the original Lick staff only James E. Keeler had any real graduate training, one year divided between Heidelberg and Berlin. Many American scientists of his generation did their graduate work in Germany, then the world center of science. But the situation was rapidly changing when Lick Observatory opened its doors in 1888, and graduate training was under way at several universities in the United States, led by Johns Hopkins, America's first research university.

Director Edward S. Holden wanted to have a graduate program at Lick Observatory from the very beginning. He was the only one of the original five staff astronomers who had a doctor's degree—not one, but two, both honorary LL.D. degrees. One was bestowed on him by the University of Wisconsin in 1886, just after he left its Washburn Observatory, and the other by Columbia University in 1887, while he was president of the University of California. Later, James E. Keeler and John M. Schaeberle both received honorary degrees from the University of California, Keeler an Sc.D. in 1893 and Schaeberle an LL.D. in 1898.

Lick Observatory's first graduate student was Armin O. Leuschner.

He arrived at Mount Hamilton fresh from the University of Michigan in late September 1888, just three months after the new observatory had gone into operation. Then twenty years old, Leuschner had been born in Detroit of German emigrant parents. His father died when Armin was less than a year old, and his mother took him back to Germany. There he received an excellent education at a Gymnasium (academic high school) in Kassel. After graduating from it in 1886, he returned to the United States and entered the University of Michigan. Two years was time enough for him to earn his bachelor's degree there, a comment on the relative preparation provided by German and American high schools in those years. Leuschner wished to become a professional astronomer, and it was natural for him to want to follow Schaeberle, who had been his professor at Ann Arbor, to Mount Hamilton.

The University of California had no formal graduate program in astronomy, but Holden simply advised Leuschner to come, and notified President Horace Davis in Berkeley: "I have taken it for granted that the University will not wish to refuse instruction to any student & I have written Mr. Leuschner to come out as soon as may be."

Upon Leuschner's arrival at Lick, Holden put him to work and wrote Davis that the young man "desire[d] to enter as a candidate for the Ph.D." Davis was not at all sure there should be graduate students at Lick Observatory and most of the professors at Berkeley did not want students anywhere except on their campus, but Holden insisted that Leuschner be admitted, writing the president:

> You will understand, of course, that it is pleasanter for all of us to be entirely free of teaching so as to be able to devote all our time to research. But I feel sure our best use to the University is to help to form a school of Astronomy of the highest grade which will attract students (like Leuschner) away from the European Observatories, as he has been attracted.

Davis and the Berkeley professors grumbled, but Leuschner stayed at Mount Hamilton. He paid no fees and had no fellowship; he received his board from Holden in exchange for tutoring the director's young son Ned. Leuschner took no formal courses, but assisted the faculty astronomers with their research, especially Schaeberle, whose meridian-circle observations required recording and tabulating large amounts of numerical data. Leuschner was learning by doing. During his first year as a graduate student, he accompanied Keeler and E. E. Barnard to Bartlett Springs on the first Lick eclipse expedition.

By the end of Leuschner's first year, a graduate committee had been appointed to guide his studies. Its chairman was Holden, and the other members were John Le Conte, the elderly professor of physics, and Irving Stringham, the professor of mathematics, representing the two "minor" fields for which Leuschner was registered. He was to study a long list of books, chiefly in celestial mechanics, but including Heinrich Kayser's textbook on spectral analysis and another book on theoretical optics. Leuschner's thesis topic, suggested by Holden, was to be on astronomical photometry. One of the main objects was to secure accurate measurements of the absolute brightness of the sun's corona, both from photographic plates taken at eclipses and from visual observations such as Leuschner had already made. The committee laid out a program that called for Leuschner to spend the fall semester of 1889 at Berkeley taking classes, all of 1890 at Mount Hamilton observing and analyzing his data, and the second semester of the 1890–1891 academic year at Berkeley, completing his thesis.

Leuschner was soon settled in Berkeley and enjoying it much more than Mount Hamilton. He liked taking the regular courses in advanced mathematics and being involved in campus life. He returned to Mount Hamilton after the end of the first semester, and continued his studies, thesis work, and practical training. In the spring and summer of 1890 he assisted Keeler at the 36-inch refractor in making his accurate measurements of the wavelengths of the strongest emission lines in the spectra of planetary nebulae. Leuschner liked Keeler and enjoyed this work, but he did not wish to become a spectroscopist himself.

That fall, encouraged by Stringham, Leuschner returned to Berkeley instead of remaining on Mount Hamilton. There was a shortage of teachers in mathematics, and Leuschner was appointed an instructor to fill the gap. This position, much like a present-day teaching assistant, enabled him to earn a little money as he continued to work on his thesis. It was not going well, and in January 1891 Holden counseled him to take another year, so that he could "do full justice to himself [and] to his subject" rather than rushing to complete it by June as originally planned.

This advice coincided with Leuschner's own thinking, and he followed it gladly. During this summer vacation he traveled to Germany, stopping off at Harvard to discuss his thesis research with the photometry experts at the observatory there. In Europe he stayed at Potsdam Observatory much of the summer, working with Hermann Vogel and Julius Scheiner. Returning to Berkeley in the fall, he continued studying and teaching mathematics for another year.

In April 1892, he decided to apply for a position as an assistant professor of astronomy at the University of Michigan, his alma mater. He asked Holden to recommend him for it, writing that if he got the job he would spend his summer vacation at Mount Hamilton and com-

plete his thesis before leaving for Ann Arbor. Holden, and no doubt
Schaeberle and W. W. Campbell as well, recommended Leuschner for
the Michigan job, and he got it. He did not take it, however. By now
he had made himself nearly indispensable at Berkeley. He was also thoroughly integrated into the powerful little German community in San
Francisco. His sponsor was Ernst A. Denicke, a wealthy businessman
and the president of the Mechanics Institute in San Francisco; through
him Leuschner had become acquainted with Regent George Kiefer. They
naturally wanted the young instructor to stay in California. The Michigan offer validated his credentials for appointment as a regular faculty
member. President Martin Kellogg, who had succeeded Horace Davis,
could recognize a bright, energetic go-getter with important supporters.
After a series of conferences with several of the senior Berkeley professors, Kellogg offered Leuschner an assistant professorship in mathematics, with the understanding that he would gradually take over the
teaching in astronomy from elderly Frederic Soulé, the engineering professor who had been doing it. Leuschner stayed.

From that moment on, relations between Holden and Leuschner began to deteriorate rapidly. The director of Lick Observatory did not
want any rival center of astronomical knowledge on the University of
California scene. Leuschner had come to realize that Holden was in-
capable in research, and did not really understand the thesis that the
younger man was supposedly doing under his direction. Like many of
Holden's projects, the idea of this research program on photometry was
a good one. It was based on his wide reading and correspondence with
leading scientists. It was exactly the field in which Edward C. Pickering
and his staff at Harvard College Observatory were making great strides
at that very time. But Holden had not really done any work in photometry himself, and that was a hopeless position from which to attempt
to direct a thesis. Furthermore, Holden could not resist posing as an
expert in the field. He issued grandiose written plans for Leuschner's
thesis, but in fact they were nothing but lists of goals and ideas, with
no concrete suggestions for carrying them out. All the available evidence
suggests that Leuschner had become more and more depressed over his
thesis, and blamed himself for his lack of progress with it until his trip
to Cambridge and Potsdam in the summer of 1891. After his conversations with Pickering, Vogel, and Scheiner, he no doubt realized that
Holden was the principal problem.

In Leuschner's first year as a regular faculty member in Berkeley, he
and Holden became embroiled in an undignified squabble. It turned
upon a visit of Leuschner's astronomy class to Mount Hamilton.
Through a breakdown in communication, he and his students did not
get to see a spectroscope on the telescope as he had promised them,

but instead were subjected to what they interpreted as a demeaning lecture by Holden. Word of this altercation got into the Oakland and San Francisco newspapers. Leuschner and Holden, each prickly individuals with a strong sense of self-importance, were both affronted. It was only the first of a series of quarrels that effectively destroyed any chance there might have been for close academic cooperation between Berkeley and Lick as long as Holden was in charge on Mount Hamilton.

The director made a few halfhearted attempts to get the Berkeley assistant professor to resume work on his thesis, but Leuschner was always "too busy" to consider doing so. Of course he did not want to put himself back under his rival's supervision. In 1896 Leuschner married Ida Denicke, the only daughter of the San Francisco millionaire who had befriended him. She was young, attractive, an accomplished musician and linguist, a blue-eyed, brown-haired "California girl" of the time who enjoyed sports and loved swimming. Their marriage occurred just before Leuschner's leave of absence for a year, which they spent in Berlin. There he received credit for his previous graduate work at Lick and Berkeley, and in one year he completed an entirely new thesis on the determination of the orbits of comets, his first love. In the "propositions" that he was prepared to defend before the faculty of the University of Berlin as part of his formal final examination, he could not resist a shot at Holden for the original thesis topic that he had recommended. Leuschner listed two serious statements, claims having to do with the superiority of the methods he advocated for calculating orbits from observational data, but his final "claim," traditionally a sort of academic joke, was that photographic measurements of the brightness of the corona could only be carried out sometime in the indefinite future, after the laws that govern the blackening of photographic plates had been established.

Leuschner received his Ph.D. from Berlin *"testimonium acuminis insigne"* (in testimony of his signal acumen), the highest degree of honor. He returned to Berkeley in the fall of 1897 just as his father-in-law was participating actively in the successful campaign to discredit Holden and drive him out of the Lick directorship.

Campbell was the first "special student" who came to Mount Hamilton, in the summer of 1890. This label designated anyone who was not a candidate for a degree. Most of them, like Campbell, were faculty members at other institutions, or teachers, who wanted to spend a summer working as volunteers at the famous Lick Observatory. As long as Holden was willing to accept them, the Berkeley administration and faculty never put obstacles in the way of the special students. Some of the other ones who arrived soon after Campbell were Henry C. Lord, who was an instructor at Ohio State University; Daniel W. Murphy, a graduate student and instructor at Stanford; Susan J. Cunningham, the professor of mathematics and astronomy at Swarthmore College; Mary E. Edwards, a graduate of Mount Holyoke and a high school teacher in New Britain, Connecticut; and Herman D. Stearns, another Stanford instructor. Of the eventual members of the Lick staff, in addition to Campbell, William J. Hussey first came to Mount Hamilton as a special student in 1893, Robert G. Aitken in 1894, and Heber D. Curtis in 1897.

The men were provided rooms in the Brick House or one of the cottages and ate their meals in the bachelor's mess that had originally been started by Schaeberle and Keeler. In the early days it usually had a Chinese cook. This communal arrangement was unthinkable for women at the time, and they had to lodge instead with one of the astronomers. Cunningham, as a senior astronomer and director of the Swarthmore College Observatory, had a separate room and took her meals with Holden during her stay, while Edwards, a younger woman, roomed with Campbell and his wife in their house on Ptolemy Ridge.

The first Lick graduate student with a fellowship was Sidney D. Townley. A native of Waukesha, Wisconsin he had attended the state university in Madison, where he studied astronomy under George C. Comstock. In his junior year Townley worked as a part-time assistant at Washburn Observatory helping his professor observe with the 15½-inch refractor. For this Townley received 20 cents an hour, and the privilege of renting a bedroom in the observatory for $4 a month. As a senior he was promoted to the time service, which provided accurate time signals for Madison and surrounding area. In this job he earned $30 a month, which enabled him to save a little money.

After receiving his B.S. in 1890, Townley decided to stay on as a graduate student and instructor. In addition to operating the time service, he taught a section or two of a beginning mathematics course each term. Now that he was an instructor he was permitted to stable the family horse in the observatory barn, whenever his father would let him ride it from Waukesha to Madison.

In the summer of 1891, after his first years as a graduate student, Townley went on a vacation trip to the west coast. He spent most of July and August with his brother Bill, who lived in Sanger, Oregon, but before going there Townley visited Lick Observatory, Berkeley, and Stanford University, then just about to open its doors to its students. Comstock, who had worked under Holden at Wisconsin, had given Townley a letter of introduction to the Lick director. Holden took him around to meet all the staff, showed him through the observatory, and invited him to dinner and to stay the evening. Campbell was working with the spectograph on the 36-inch that night, and Barnard was mak-

ing visual observations with the 12-inch. The next day Townley went
down to San Jose, and the following day he visited the Stanford campus,
where he admired the handsome new buildings and opined, "This is
the beginning of a fine University and will probably some day rank
among the finest in the land." The following day he went on to Berkeley, where an old friend from Waukesha, "Pancake" Shaw, guided him
around the university. Here Townley was not as impressed, or perhaps
he was a victim of local pride, for he recorded his judgement that "the
University has a nice location and some fine buildings but in neither
respect does it come up to the U[niversity of] W[isconsin]."

During his second year as a graduate student at Madison, Townley
applied to both Harvard and Lick for fellowships to work for the Ph.D.
degree, which he could not get in astronomy at the University of Wisconsin. Comstock recommended him strongly to both institutions.
Townley did not get the Harvard fellowship, but Holden awarded him
the first Phoebe Apperson Hearst Fellowship, with funds he had just
obtained from the wealthy widow of Senator George Randolph Hearst.
The fellowship was worth $360 for the year, and in addition the student
got an unfurnished room for working on the time service. Holden assured Townley that "no doubt" he could renew it for a second year.

Townley arrived at Mount Hamilton in mid-July to begin his fellowship work. In addition to the time service, he assisted Campbell with
his spectroscopic work with the 36-inch refractor, and fixed up a small
telescope for his own photographic measurements of variable stars.
Lord and Murphy, from Ohio State and Stanford, were both special
students that summer, and Townley enjoyed their companionship. They
returned to their universities in September, leaving the young Hearst
Fellow the only student on the mountain. Although the Lick astronomers were friendly to him and he enjoyed his work, he felt increasingly
isolated. Thus he was glad to go down to the Berkeley campus for the
second semester. He took courses in differential equations, functions of
a real variable, and mechanical quadratures, the first two taught by
Stringham and the last by Leuschner. In addition Townley did some
experiments on his own with lenses and telescopes in the physics laboratory.

However, his career took a sudden turn when Holden committed all
the money he had received from Hearst to Schaeberle's eclipse expedition to Chile, and had nothing to offer Townley to continue his fellowship into 1893–1894. The director "hoped" that something would
turn up, but was far less definite than he had been the previous spring
when he had been persuading the first Hearst Fellow to come to California. Townley quickly began exploring his other options, and managed to land a job as an instructor in astronomy at the University of

Michigan. He left Lick Observatory after only one year, and there never was another Hearst Fellow.

Townley continued his graduate work at Michigan for two years, then spent one year studying in Germany. After one more year at Ann Arbor, he received his Sc.D. in 1897. He had been a graduate student for a total of seven years at Wisconsin, California, Michigan, Berlin, and Munich. After a few years at Berkeley, teaching under Leuschner, and then a few more observing at the International Latitude Station at Ukiah, California, Townley was appointed to the Stanford faculty in 1907. He taught astronomy there until he retired in 1932, taking a very active part in the Astronomical Society of the Pacific and returning to the Lick Observatory on many visits with his classes. Townley died in 1946, just as Stanford was beginning to transform itself into the fine university, ranking among the first in the land, that he had foreseen on his first visit, more than half a century before.

After this experience with the first, last, and only Hearst Fellow, Holden realized that Lick Observatory needed a firmer base for graduate student support than the year-to-year whims of a wealthy widow. He proposed a fellowship program to the regents to be funded by outside donors for whom the fellowships would be named. A gift of at least $1,500 would be required, to support one student for three years, and the preferred amount would be $10,000, which would provide enough income to finance the fellowship "forever." The regents were glad to approve the concept, but none of them came forward as donors, nor did Holden succeed in finding any other wealthy individuals who would do so.

In 1897, however, the legislature put $480 per year into the Lick Observatory budget earmarked for one fellowship. Lord, by now professor of astronomy at Ohio State University, recommended his prize undergraduate student, Ernest F. Coddington, for it. The young Buckeye got the fellowship and arrived at Mount Hamilton on August 1, 1897, just as the final episodes in the drama of Holden's forced resignation were working themselves out. When Holden finally departed on September 18, 1897, he had not produced one Ph.D. despite all his efforts in his nine years as director. Part of the problem was that he could not do first-class astronomical research himself, and was therefore incapable of directing graduate students in their theses. His attempts to do so, in the cases of Leuschner and to a lesser extent Townley, only awakened them to his shortcomings. Even more importantly, his implacable determination to retain complete control of the astronomy students antagonized the Berkeley professors. Leuschner knew all too well how to make the most of this constant source of tension.

The whole situation changed dramatically when Keeler arrived back at Mount Hamilton as the second director of Lick Observatory. From his student days at Johns Hopkins, where he and the other undergraduates were outnumbered by the graduate students, and his year of graduate work at Heidelberg and Berlin, he knew the importance of a firm contact between research and teaching. Keeler was determined to have a sound, productive graduate program at Lick. He was able to do something about it, too. There was one open position on the Lick staff, the result of the resignation of John M. Schaeberle. Keeler was besieged by suggestions as to whom to appoint to fill this job. Townley was one of the prime candidates for it. But instead of appointing any new faculty astronomer, the new director decided to leave the job open and to use the money thus saved for three graduate fellowships. This was the origin of the Lick Observatory Fellows program.

As soon as he arrived in California in May, Keeler revealed his plan to Leuschner, who as a graduate student had worked under him not only at Mount Hamilton, but also for one summer at Allegheny Observatory in 1893. Leuschner, who trusted Keeler as he had not trusted Holden, was all for the plan. In one stroke Keeler gained two allies, not only the head of the Berkeley astronomy group, but also his father-in-law, the powerful Regent Denicke. Keeler next sold the plan to Timothy Guy Phelps, the wily old politician who was chairman of the regents' Lick Observatory Committee. With these two regents in his corner, Keeler had no trouble convincing the rest of them that his fellowship plan was a sound one. It went into operation that summer.

Part of the idea of this plan was that University of California graduates should have first preference for the fellowships, if their training and abilities were as good as those of any outside candidates. That had sweetened it for Leuschner. All three of the first Lick Observatory fellows, Russell Tracy Crawford, Harold K. Palmer, and Frank E. Ross, were Berkeley graduates. Each of them received $600 a year, a good sum for a fellowship in those days. Coddington, whose fellowship under the regular funds was continued, got the same amount. They were all expected to spend the summer and fall at Mount Hamilton, assisting the professors and doing their own thesis research, and to be in residence in Berkeley to take courses in the winter and spring, the poor observing season in California. Both Leuschner and Keeler had to approve each student's program.

Crawford had completed his undergraduate work at Berkeley in 1897, and had done one year of graduate study on the campus with Leuschner before the Lick program began. He wanted to start right to work on a thesis. Keeler suggested that he do an observational determination of the atmospheric refraction at Mount Hamilton, partly based on existing observational data that had been taken by Richard H. Tucker of the Lick staff, and partly on new measurements that Crawford would make himself. Crawford did this thesis under Tucker's supervision. He worked on it for three years. He spent the second semester of each academic year in Berkeley, taking courses in mathematics, theoretical astronomy (orbit theory), and a little physics (mostly mathematical) in 1899 and 1900, and writing and correcting his thesis in 1901. He passed his final examination with flying colors on May 3, 1901, and received the first Ph.D. in astronomy granted by the University of California. Keeler was dead by then, but the program that he had founded was beginning to produce.

Ross, the second of the first three Lick Observatory Fellows, was a native San Franciscan, who had completed his undergraduate work at Berkeley in 1896. Then he had taught mathematics and physics for a year at Tamalpais Military Academy, a prep school in San Rafael, before returning to the campus as a graduate student in mathematics in 1897. He had dropped out and taken a job with the Coast Survey the next summer, but was rescued by the Lick Fellowship program just a few months later. At Mount Hamilton he assisted William J. Hussey, reducing his double-star observations. Ross had severe eye problems, together with a strong tendency (that lasted all his life) to want to leave whatever job he had for what at the time seemed to him a better one that he did not have. After only one year he decided that observational astronomy was too hard on his eyes, and he switched back to mathematics. He completed his thesis on differential equations and received his Ph.D. in mathematics at the same ceremony at which Crawford got his degree in astronomy. Ross then became an assistant professor of mathematics at the University of Nevada for a year. Next he went to Washington where he worked as an assistant under Simon Newcomb for several years, and after several other jobs ended up on the Yerkes Observatory staff. In spite of his almost non-existent training in observational astronomy, Ross had by far the most productive research career of the first three Lick Fellows.

Palmer, the youngest of the three, had just finished his undergraduate work in Berkeley before coming to Mount Hamilton in the summer of 1898. He was assigned to work with Keeler, and assisted the director with his direct photographs of nebulae with the Crossley reflector. Palmer was a good observer, and after Keeler's death Campbell kept him at work with the Crossley, now assisting Charles D. Perrine. Each year Palmer spent one semester at Berkeley, taking graduate courses. His was an independent soul, as shown for instance by his report from the campus to Campbell in September 1901:

My work here this year is not so bad as it has been before as I managed to escape all work with Leuschner, still it is bad enough. I am taking Elliptic Functions, Theory of Functions, and work in the physics laboratory on the Zeeman effect.

He went on to say that he was still supposed to take courses for his Ph.D. in hydrodynamics, electricity, and heat as a form of energy, "but I have no time for them and no one here objects to my leaving them out entirely if you do not." Campbell, normally a stern disciplinarian, accepted it because he knew Palmer was good, and because he saw so much of Keeler in him.

Campbell set Palmer to work on a thesis with the Crossley reflector, using an improved version of the small slitless spectrograph that Keeler had designed and had been ready to put into use just before his death. However, Palmer lacked Keeler's drive and this project drifted on for years. Campbell had hand-picked Palmer to go to Chile with William H. Wright, to assist him at the planned Mills southern observing station. As the telescope for this observatory was delayed, Palmer became increasingly disenchanted, particularly as his fellowship stipend had not been raised as he had hoped. He complained that he was "practically busted" and that there was no way he could save any money on his salary in Berkeley. He did not like the idea of the Chile expedition, but said that as he had been too lazy to find any other job he would go, but only if he were paid $900 for the first year ("which means that I will go for $900.00 but not for $899.00 or less") and that "in addition . . . I shall want a written guarantee signed by the President that my salary for the second year shall not be less than $900 under any circumstances."

In the summer of 1902 Palmer worked with the Crossley nearly every dark night, securing low-dispersion spectra with the slitless instrument of many planetary nebulae, a few of their central stars, and several novae ("new stars"). It was exactly the program Keeler would have done if he had not died prematurely. These spectra formed the material for Palmer's thesis, in which he described the ultraviolet spectra of planetary nebulae, including the two new emission lines he discovered, now known to result from highly ionized neon, [Ne V]. He had no job and went to Claremont, in southern California, to stay with relatives for a time. He was still waiting to go to Chile, and wrote Campbell sarcastically "I hope the mirror won't break while I am [here], for I should hate to be stranded in [this] place." In 1903 a temporary instructor's job opened up in Berkeley, and Leuschner gave it to Palmer.

Then suddenly the telescope was completed, and the expedition was about to get under way. Leuschner and Campbell pressed Palmer to take his final examination for the Ph.D. degree. He had completed his thesis, and it was ready to be published, but he wanted to put off the exam until after he had returned from Chile. Leuschner practically dragged him to the examination on February 27, 1903, the day before he and Wright sailed for Valparaiso. Though he had done his best to avoid the examination, afterward Palmer was able to write Campbell:

> It is all over at last and after firing questions at me for over an hour and a half the committee decided to give me my degree. I managed to get through without getting very badly scared. As the examination was mostly on the work I did at the L[ick] O[bservatory] it was easy.

He thus became the second Ph.D. in astronomy from the University of California.

In Chile Palmer did an excellent job assisting Wright, as he had previously done for Keeler on Mount Hamilton. But the young assistant complained increasingly of the long hours of night work at the telescope. He was convinced that he could not keep it up. Campbell had essentially promised to give him a job on the Lick faculty after he had completed his term at Santiago, but Palmer did not want it when he came back at the end of 1905, because it would mean he would have to go back to nighttime observing. He tried to get a job in Los Angeles, but when it fell through he accepted an appointment to measure spectrograms at Mount Hamilton. It was a daytime job, but only a temporary one. After a few months he resigned to take a post as an assistant at the new Mount Wilson Solar Observatory, which George Ellery Hale was just getting started in southern California. Solar observing was daytime work, and Campbell had recommended Palmer to Hale for the job. But after only one summer the loneliness of life on Mount Wilson got the best of Palmer. He quit astronomy and became an irrigation and sanitary engineer, and worked on many large projects in the Los Angeles area. But even forty-five years later, when he revisited Lick Observatory and the scenes of his youth, he still had "a soft spot in my heart for Mt. Hamilton, which I never had for Mt. Wilson."

Coddington, the one fellow who had been at Mount Hamilton before Keeler came, bringing Crawford, Ross, and Palmer with him, did not complete his Ph.D. at the University of California. He had also assisted Keeler at the Crossley reflector, and Hussey with his double-star measurements. But in 1900 Coddington had an opportunity, thanks to a wealthy supporter of his alma mater, Ohio State University, to go to Germany for further graduate work. It was still the world center of science, and Coddington seized his chance. He left for Berlin, where he received credit for his work at Lick, and completed the requirements for his Ph.D. He returned to America and a long career as professor of mathematics and geodesy at Ohio State.

Probably the all-time greatest researcher produced by the Lick graduate program was its third Ph.D., Joel Stebbins. Born in Omaha, he attended the University of Nebraska, and after graduating in 1899 he stayed on for one year of graduate work. He accompanied his teacher G. D. Swezey east to see the solar eclipse in the summer of 1900, just as his contemporary outstanding researcher-to-be, Henry Norris Russell, than a student at Princeton, accompanied his teacher Charles A. Young south to the same event. Stebbins and Swezey visited the Lick eclipse camp at Thomaston, and met Campbell there. Then Stebbins spent one year at the University of Wisconsin as a graduate student with Comstock, but the young Cornhusker was clearly destined for bigger things. Comstock advised him to move on either to Lick Observatory or to Yerkes, where he would be trained in the "new astronomy," astrophysics. Stebbins applied to both, but Hale, then still director of Yerkes Observatory, had no funds to offer him, while Campbell could give him a Lick Fellowship at $600 a year. Stebbins went to California.

He went straight to work, observing with the Crossley reflector his first semester. Stebbins enjoyed the life on Mount Hamilton and soon was writing his parents with delight of how he and the other students had taken the new observatory secretary, Benjamin F. Mills of Boston, out to "bag quail" one dark night, and lost him in a dense thicket at the other end of the mountain. It was the California version of a snipe hunt. Stebbins spent his second semester on the campus in Berkeley where he took theoretical astronomy (orbit theory), spherical harmonics (an advanced mathematics course), spectroscopy, and spectroscopic laboratory. In his spare time he was supposed to learn the equivalent of three years of Latin, a requirement for a University of California Ph.D. but not at Nebraska. Luckily for astronomy, the examiners for this requirement were tolerant of midwestern science students.

In May the courses ended and Stebbins returned to Mount Hamilton. The rainy season had just ended and the mountains were green, covered in many areas with lupines and orange California poppies. It was beautiful. But as summer progressed everything dried out and the hills turned brown. Stebbins' thoughts returned to Wisconsin, and he wrote his parents that "I can't make up my mind to think that this place is as pretty as Madison. Everything is so different." Campbell put Stebbins to work on a thesis with the 36-inch refractor, taking spectrograms of the long-period variable star o Ceti as it gradually became fainter, and then brightened up again. Stebbins' description of the changes in its spectrum was a classic in the study of variable stars. Full of drive and eager to finish his degree, Stebbins took his final examination in May 1903, just a few months after Palmer, though he had come to Lick Observatory three years later.

Stebbins then got a faculty job at the University of Illinois on the recommendation of Campbell. There was only a small telescope there, but in the physics department Stebbins met Jakob Kunz, who was making and experimenting with primitive photoelectric cells. These cells convert light falling upon them into a weak electrical current. Stebbins recognized their great potentialities for astronomical photometry. With a photoelectric cell placed at the focus of even a small telescope, he could accurately measure the light from a star by measuring the electrical current, or the charge on a galvanometer, that it produced. In his first experiment the photocell that Kunz had made was so insensitive that Stebbins could not measure even the brightest star with it. Always creative, Stebbins turned it on the moon, which was bright enough. He measured accurately, for the first time, the brightness of the moon as a function of phase, or in non-technical terms, how effectively it reflects and scatters sunlight in various directions.

Both Stebbins and Kunz rapidly improved their techniques. Stebbins used photocells and later photo-multiplier tubes to measure the magnitude of stars, clusters, and galaxies. With these measurements he made many important contributions to understanding pulsating stars, stellar temperatures, interstellar matter, the structure of our Galaxy, and even the make-up of distant galaxies. Along the way he returned to the University of Wisconsin as professor of astronomy in 1923, when Comstock retired. Stebbins, a great raconteur, was a favorite after-dinner speaker at astronomical meetings for thirty years or more. Many of his stories drew heavily on his experiences in the early days on Mount Hamilton. After his own retirement in 1948, he moved to San Jose, to be near Lick Observatory again. He visited the observatory weekly, and participated actively in its research for many years before his death in 1966.

Another outstanding research worker produced by the Lick graduate program was Stebbins' friend and fellow student Ralph H. Curtiss. A native of Connecticut, Curtiss did his undergraduate work at Berkeley, and after graduating in 1901 he stayed on as a graduate student. He took many of the same advanced courses as Stebbins. Curtiss did his thesis, also under Campbell, on spectroscopic observations of the Cepheid variable W Sagittari with the 36-inch refractor. In accordance with the ideas of the time, he interpreted the radial velocities he measured as if the star were a spectroscopic binary, or double-star system with one member much brighter than the other. His solution showed the two stars would be very close to one another, and he attributed the light variations to the large tidal forces one of them would necessarily exert on the other. This was the accepted thinking of the time, later shown to be incorrect. Lick theses under Campbell were strong on hard observational data, but were not noted for conceptual breakthroughs.

After receiving his Ph.D. Curtiss spent a few years at Allegheny Observatory, and then went on to the University of Michigan, where he built up a noted spectroscopic observing program, modelled on his Lick experience. At Ann Arbor, Curtiss and his collaborators confirmed observationally many of the predictions of the new theory of Cepheid variables, put forward by Arthur Eddington, which interpreted stars like W Sagittari as pulsating single stars rather than orbiting double stars. Lick graduates could keep up with the times. Curtiss succeeded Hussey as director of the University of Michigan Observatory, but died of a heart attack in 1929 at the age of forty-nine, ending a short but very fruitful observational research career.

The Lick student whose thesis contained the most important new observational result was Edward A. Fath, though it went unrecognized for many years. In fact, he hardly seemed to recognize it himself. Fath came to Lick from Carleton College, where he had earned his B.S. degree in 1902. For his thesis research he studied the spectra of spiral "nebulae," beginning his observational work with the small quartz spectrograph that Keeler had designed for the Crossley reflector, and that Palmer had used for his thesis. The physical nature of the spirals was an unsolved puzzle at the time. The spectra did not have emission-line spectra, showing that they were not hot gas clouds like planetary and diffuse nebulae. Instead the spectra of the spirals seemed continuous, like the spectrum of a heated solid or liquid object, such as a block of hot steel or a ladle of molten iron. Yet the spirals had such faint surface brightnesses that they could not be solid or liquid bodies. Scheiner in Germany and William Huggins in England had reported seeing weak absorption lines, similar to those in stellar spectra, in the spectrum of M 31, the Andromeda nebula. Their spectrograms were so faint, however, that it was not at all clear if these features were real or not. Huggins also reported seeing emission lines, making the puzzle even more confusing. Keeler had designed the fast spectrograph for the Crossley specifically to solve this problem, but he had died before he could start using the instrument. Campbell had put it aside, except for Palmer's thesis, to concentrate the observatory's resources on the radial-velocity program, of which Stebbins' and Curtiss' theses formed parts.

Fath made some preliminary trials with the quartz spectrograph. The results were promising, but showed that he would need a higher-dispersion system to see if absorption lines were indeed present. Campbell allowed Fath to have a special spectrograph constructed for observing very faint objects, such as the spiral nebulae. It was built around a single glass prism, and used a pair of lenses to give a dispersion of 400 Å/ mm, just sufficient to see the absorption lines if they were there. The lenses provided an effective focal ratio of F/3, very fast for the time.

With this spectrograph on the Crossley reflector, Fath took photographic exposures of the Andromeda nebula extending over several nights. One plate had a total exposure of 8 hours and 47 minutes; a second, 18 hours and 11 minutes. Both these spectrograms showed definite absorption lines, at the wavelengths of the absorption lines in the spectrum of the sun and similar stars. Yet the Andromeda nebula had a much fainter surface brightness than the sun. Fath made the only possible interpretation. The Andromeda nebula must be a very distant "star cluster" (or galaxy, as we would say today) composed of very many stars, giving the solar-type absorption-line spectrum, but with much empty space between them, resulting in the very low surface brightness.

To check his reasoning Fath took spectrograms with the same instrument on the same telescope of globular clusters, groups containing numerous faint stars that could be seen individually on long-exposure photographs. No one could doubt that they were star clusters. Their spectra, Fath found, were also continuous with the same absorption lines, just as in the Andromeda nebula. He had proved spectroscopically that it was composed of stars. Fath obtained spectrograms of several other spiral "nebulae." Some were so faint that even 8- or 14-hour exposures were too short to reveal any absorption lines. A few showed bright emission lines. These were peculiar objects, Seyfert galaxies that contain, in addition to stars, copious amounts of ionized gas, but they confused Fath and made him distrust his main result. A few other spirals showed absorption lines like the sun, but not as many as in the spectrum of the Andromeda nebula. These spectrograms were not as well exposed as those of M 31 itself.

Fath published his thesis in 1909. He clearly stated his result that M 31 in particular, and the other spirals in general, must be very distant and must consist of large numbers of individually faint stars, like globular clusters. He weakened his paper by stating his conclusion would stand or fall on measurements of parallax (that is, of the geometrical distance) of the spirals. No doubt the careful, conservative Campbell advised him to include this reservation. As a result of it, however, equally conservative astronomers were able to ignore Fath's result. It was only when Edwin Hubble convincingly identified individual Cepheid variable stars in M 31 with the 100-inch Mount Wilson telescope in the 1920s that spiral "nebulae" were recognized as *galaxies* by all astronomers.

Fath himself worked three years at Mount Wilson, and further confirmed his Lick result, but no one really listened. He moved on to a teaching job in the Midwest, and when he returned to Lick on a year's sabbatical leave in 1933–1934, he worked not on galaxies but on the complex, pulsating variable star δ Scuti. When he died in 1959, the Lick astronomers considered that this had been "perhaps his most significant work."

Three of the early staff members at Mount Wilson Observatory were graduates of the Lick Ph.D. program. They were Paul W. Merrill, who finished in 1913, Seth B. Nicholson, in 1915, and Roscoe F. Sanford, in 1917. Merrill's thesis was a spectroscopic study of hot stars with bright hydrogen emission lines in their spectra, indicating that they have shells or envelopes of hot gases surrounding them. He continued to work on the spectra of these stars and of long-period variable stars for the rest of his long and highly productive career.

Nicholson, while he was still a graduate student at Berkeley in 1914, discovered a previously unknown satellite of Jupiter, following in the footsteps of E. E. Barnard and Charles D. Perrine. Nicholson had been sent to Mount Hamilton to take direct photographs with the Crossley reflector for measuring the positions of the known satellites of the giant planet, to check and improve their orbital data. On a pair of photographs of the outermost "eighth satellite" Jupiter VIII he had taken on July 21 and 22, Nicholson detected what appeared to be another, much fainter moon, also moving with the planet. He followed up with check photographs on the next several nights, which confirmed it as Jupiter IX. Nicholson's Ph.D. thesis was the calculation of the accurate orbit of this moon of Jupiter that he had discovered himself.

Hired on the Mount Wilson staff after he had finished his degree, Nicholson was put to work on solar observational research. Most of his career was spent on this program, but his own deepest interests were in the orbits of satellites, asteroids, and comets. With the 60- and 100-inch telescopes, he discovered one after another three more satellites, Jupiter X, XI, and XII, equalling the record of Galileo, who had discovered the first four Jovian satellites three centuries before.

Sanford, who had finished his undergraduate work at the University of Minnesota in 1905, started his astronomical career as an assistant to Tucker at Lick Observatory, making precise stellar position measurements with a meridian circle. In 1908 Sanford accompanied Tucker to Argentina for two more years of this work in the Southern Hemisphere. Then Sanford took a job as an assistant at the Lick southern station in Chile, obtaining spectrograms of stars and measuring their radial velocities. When he returned to the United States in 1915, he had almost ten years experience in routine measuring programs. He received a fellowship at Lick and in two years did an excellent thesis, based on the ideas of Heber D. Curtis, which proved very clearly that the Milky Way system is a spiral galaxy. Sanford used observational data on the distribution of spiral nebulae in space, as indicated by their positions in the sky, their magnitudes, and their apparent angular sizes, to show that they must lie outside the Milky Way. Their high radial velocities, measured spectroscopically by the Doppler effect, also proved this, he said. Their absorption-line spectra suggested "that they are composites of great numbers of stars, the integration of whose light builds up the spectrum that we photograph. We are not able to resolve these nebulae into component stars, and, therefore, if they are composed of stars, they must be at tremendous distances," he wrote.

Sanford explained the apparent clustering of the spiral nebulae to the north and south poles of the Milky Way as resulting from "absorbing or occulting matter" in the galaxy's plane. It was just what would be expected if the Milky Way was another spiral "nebula" itself, with occulting matter (interstellar dust) concentrated in its principal plane, as in other spirals. Sanford even recognized that the yellow and red stars are more strongly concentrated to the center of "the Galaxy" (his words, and ours today), while the blue stars are more strongly concentrated toward the outer parts of the plane: a primitive statement of the idea of two stellar populations Walter Baade put forward three decades later. Sanford's thesis revealed a very clear picture of galactic structure essentially as it is understood today. Heber D. Curtis, his thesis advisor, had provided the main ideas and Sanford had fleshed them out wonderfully. He provided much of the quantitative hard data that Curtis used in the Great Debate with Shapley a few years later.

Sanford received his Ph.D. in 1917, and after a few months at Dudley Observatory in New York, joined the Mount Wilson staff. There he worked for many years on stellar spectroscopy, especially of carbon stars. He never returned to the interpretation of galactic structure to which he had made such a far-reaching contribution in his thesis.

The first women who received Ph.D. degrees at Lick Observatory were A. Estelle Glancy and E. Phoebe Waterman, both in 1913. Holden welcomed women as special students from the start, but by the time Keeler came as director the mountain was crowded, and he turned female candidates away on the grounds that there was no suitable place for them to live. Apparently the first woman to apply to come to Lick as a candidate for a Ph.D. had been Adelaide M. Hobe, who completed the full undergraduate astronomy program at Berkeley in 1899. She was a good student, especially well qualified in "theoretical astronomy" (celestial mechanics), according to Leuschner. She worked for him as an assistant, secretary, and librarian, but Keeler, though he said he otherwise would have been glad to appoint her to a fellowship, felt it would be impossible "for her to arrange for board and room on the mountain."

Another woman, Elizabeth Wylie, applied for admission in 1900, just when Keeler was looking desperately for a graduate student assistant to replace Coddington. She was a Wellesley graduate with training in mathematics and astronomy. She lived in Gardiner, Montana, where her father ran a dude ranch and pack-train outfit into Yellowstone National Park. It certainly appears that she could have survived on Mount Hamilton, and could have helped out with the observing just as much as Crawford, Palmer, and Ross had when they started two years before, but she did not get the chance. Keeler told her that the quarters on Mount Hamilton were "very limited" and "not arranged with reference to the accommodations of ladies," and advised her to try Berkeley. She dropped the idea of graduate work in astronomy. Keeler and all the early astronomers at Lick shared the bias of their times.

Campbell was more willing to make room for women if he were convinced he really needed them. After promoting Perrine from secretary to full-time astronomer, the young director had two disastrous experiences with his successors, young male secretaries who were not interested in astronomy nor, it soon developed, life on Mount Hamilton. Campbell then hired the first female secretary at Lick Observatory in 1902. She was Wilmetta Curtis, a friend of Aitken and his wife who came highly recommended by them. Curtis proved highly efficient; she could do all the work Campbell gave her and more. After two years she tired of the isolated mountain life and resigned to take a job in a law office in Oakland, but she had broken the female barrier on Mount Hamilton. From then on Campbell hired only women as secretaries at the observatory.

A few years later he hired Hobe as an assistant at Lick, but she did not get to observe. Her job was to measure the spectrograms taken by others and do the numerical computations necessary to calculate the radial velocities of the stars from these measurements. She was accurate and hardworking, and contributed greatly to the research program of the observatory. In the years between 1912 and 1922 she personally measured nearly all the spectrograms taken on the radial-velocity program. However, for her there was no chance for advancement. She knew that as a woman she could not hope to be promoted to a regular astronomer. A slight, bookish person, she suffered several apparently psychosomatic illnesses and one or more "breakdowns," probably brought on as much by the frustrations of her job as by anything else. In 1922 she left the mountain and became a statistician in the Ford Research Institute at Stanford University. Undoubtedly, if she had been a man, she would have had a far more fulfilling career in astronomy.

By 1905, when Campbell hired Hobe at Mount Hamilton, the mores of American society had changed enough so that the director could imagine women living and working at the observatory. There was more space in the dormitory, too. Both Glancy, a Wellesley graduate, and Waterman, who had bachelor's and master's degrees from Vassar, came to the University of California as graduate students soon after that.

After spending a year on Mount Hamilton, Glancy did her thesis in Berkeley, under Leuschner's supervision, on the orbits of a special type of asteroid ("the Hecuba group"). Waterman did an observational thesis at Lick on the spectra of A-type (relatively hot) stars in the red spectral region. Glancy especially was considered a very good scientist, but even though women were accepted as graduate students by 1906, it was much more difficult for them than for men to find research positions in astronomy. After getting her Ph.D. degree in 1913, she could not get a job in any North American institution. Perrine, the former Lick secretary and astronomer, hired her on the staff of the Argentine National Observatory in Cordoba, where he was director. Her work was mostly routine computing, and American scientists in Argentina, even Perrine himself, felt themselves under constant anti-foreign pressure. Glancy returned to the United States in 1918, when she observed the total solar eclipse with the Lick party at Goldendale, Washington. She ended up as a lens designer for the American Optical Company in Massachusetts. In 1922 when Campbell was casting about for an astronomer to send to Chile to take over the southern observing station in Santiago, Curtis commiserated with him that none of the men who were being considered as candidates was adequate, and added, "Too bad A. E. Glancy has not Alfred instead of Ann[a] for a first name." In other words, she would have been perfect for the job, if she were only a man.

Phoebe Waterman married soon after she got her degree; as Mrs. Otto Haas she retained a lively interest in astronomy but did no more research or teaching. There were ten women in all who earned Ph.D.s by 1931 (compared with thirty-nine men). Of these, six were married; of these six, two had jobs in astronomy while two more, Priscilla Fairfield Bok and Mary Lea Heger Shane, worked to some extent on research with their husbands, who were professional astronomers.

Of the four single women Charlotte E. Moore, who received her Ph.D. in 1931, had the most productive research career, but she never got the faculty position for which her skills, knowledge, and research ability qualified her. She had graduated from Swarthmore in 1920, and then worked for five years as a full-time research assistance for Henry Norris Russell at Princeton. She become very expert in the classification and analysis of laboratory spectra, and on the identification of spectral lines in stars. From 1925 until 1928 she worked at Mount Wilson Observatory, more or less on loan from Russell, on the identification of the tens of thousands of lines in the solar spectrum. This project was headed by Charles E. St. John, of the Mount Wilson staff, but most of the identifications were made by Moore.

In 1928 she returned to her job in Princeton. She wanted to become a graduate student and earn her doctor's degree, but Princeton accepted male students only. In 1929, Russell went abroad for over a year with his family on sabbatical leave. On his advice, and with a Lick Observatory Fellowship for which she was recommended by Aitken and Leuschner, Moore entered the University of California. She spent her first month at Mount Hamilton; then she went to Berkeley for the first semester, where she took five courses in astronomy and physics, earning As in all of them. She passed her qualifying examination "exceedingly well," and went to work in Berkeley on her thesis, treating the identification of spectral lines in sunspots, those regions on the sun cooler than the rest of its surface. This thesis had been suggested by Russell, and was based on observational material from Mount Wilson. She had made a good start on the research in Princeton, before she came to California. At the end of the second semester she returned to Mount Hamilton for the month of June. Moore then went on to Pasadena where she worked from July through December at the Mount Wilson Observatory offices, still supported by her Lick Fellowship, completing her thesis.

Russell returned from Europe in October and by November was in Pasadena himself, helping Moore with the last stages of her dissertation. She took her final examination in January 1931, completing the work for her Ph.D. in one and a half years, only two months of them spent at Lick, and only two semesters at Berkeley. One month later she was back at work at her non-faculty job in Princeton, assisting Russell.

Moore published her thesis as a long paper, plus a 45-page table of the wavelengths and identifications of the spectral lines in sunspots. In addition, she published two further compilations on the basis of her thesis work: one on the wavelengths of spectral lines in the sun and stars (*A Multiplet Table of Astrophysical Interest*), the other on atomic energy levels (*Term Designations for Excitation Potentials*). It was certainly the most productive Lick Observatory Fellowship ever awarded. If Moore had been a man, she would certainly have been hired on the Mount Wilson staff. Instead, she continued to work with Russell at Princeton until 1945, when he retired. During that period she prepared the *Revised Multiplet Table,* used by a generation of astronomers to interpret the spectra of stars. After Russell's retirement, Moore left Princeton and became a research physicist at the National Bureau of Standards. There she headed the compilation of the *Atomic Energy Lev-*

els, a series of volumes that have proven even more useful than the *RMT.*

By 1931 the University of California graduate program had turned out forty-nine Ph.D.s in astronomy. Nearly all the best of them had been Lick Observatory Fellows. Outstanding undergraduates, they had come to California from all over the country. It was widely regarded as the most distinguished astronomical graduate school in the country. As Leuschner wrote while still the chairman of the Berkeley astronomy department, it had produced more observatory directors, more members of the Astronomy Section of the National Academy of Sciences, and more outstanding research astronomers than any other. Very shortly, some of the very best of the Lick Ph.D.s began to be hired by Lick Observatory itself.

11

Double-star Observer

1923–1935

Robert G. Aitken came to Lick Observatory as a young faculty member in 1895, and stayed for forty years. As associate director, he was in immediate charge on Mount Hamilton in the years from 1923 through 1930, while W. W. Campbell served in Berkeley as president of the University of California, but retained the directorship of Lick Observatory. In 1930 when Campbell retired, Aitken became the director in name as well as fact.

Aitken was born at Jackson, California, in the Gold Country east of Sacramento, in 1862, and was only two years younger than Campbell. Aitken's parents sent him east to Williams College, in Massachusetts. It is the school whose educational philosophy was epitomized as "Mark Hopkins on one end of a log, and the student on the other." Hopkins was the president while Aitken was enrolled at Williams, but he was more strongly influenced by Truman Safford, the professor of astronomy and an assiduous double-star observer.

After his graduation in 1887 Aitken returned to California, married his sweetheart Jessie Thomas, and got a job as a schoolteacher in Livermore. In 1891 he became an instructor in mathematics at the University of the Pacific, a little Methodist Episcopal college located between San Jose and Santa Clara, a few miles from Mount Hamilton. Aitken stayed there four years, but his heart was in astronomy. He made it a point to get to know Edward S. Holden, the Lick director, and spent a few weeks at Mount Hamilton as a special student in the summer of 1894. By the next year Aitken had decided to make his career in astronomy, and when Holden offered him a job as an assistant at a quite modest salary, Aitken snapped it up. A few months later Holden promoted Aitken to assistant astronomer and he remained at Lick Observatory until his retirement.

Aitken's observing program from the start was the visual measurement of double stars. S. W. Burnham had begun this work at Mount Hamilton during his early site survey, long before the observatory was built, and had continued it with the 36-inch refractor when he returned as one of the original staff members. After his departure in 1892, there had been no regular double-star program at Lick until Aitken joined the staff.

Visual double-star observing requires keen eyesight, and also great patience, powers of concentration, and perseverance. Aitken possessed them all in abundance. The observer from time to time must measure and record the angular separation and position angle (orientation) of the pair of stars that make up a binary system whose typical periods may be tens or hundreds or even thousands of years. Only after an appreciable fraction of the period do the form and dimensions of the orbit become apparent. From them, if the distance to the binary can also be determined, the masses of the individual stars can be calculated. This is the fundamental numerical information that is the goal of double-star astronomy.

Aitken began by measuring double stars from lists of previously known pairs, but he soon became convinced that the most important project he could carry out would be a systematic survey of all the stars down to a well-defined limit of brightness, to see what fraction of them were binaries. He began this survey in 1899. The limit he chose, well matched to the 36-inch refractor, was ninth magnitude. William J. Hussey, Campbell's friend who had been appointed to the Lick staff in 1896, joined with Aitken in this program a few weeks after its start. They divided the sky into alternate zones, which they observed individually. In 1905, when Hussey left Mount Hamilton to become director of the University of Michigan's observatory, Aitken took over his zones and completed the program alone. By the time he completed the survey in 1915, Aitken had found 3,100 "new" (previously unknown) pairs, in addition to the 1,300 that Hussey had discovered, Nearly all of them were "close pairs" with separations smaller than 5".0. Aitken found many especially interesting individual systems, and determined several important general statistical results. Perhaps the most interesting of these was that at least one of every eighteen stars as bright as ninth magnitude is actually a close double star that can be "resolved" (seen as two stars) with the Lick refractor.

Aitken's monumental contribution to astronomy was his *New General Catalogue of Double Stars within 120° of the North Pole,* published in two large volumes in 1932. It was, as its name implied, a sequel to

Burnham's *General Catalogue of Double Stars* published in 1906. These catalogues gave, for each binary listed, all the published measurements up to the date of compilation, as well as many unpublished ones from manuscript records. They thus included not only their authors' measurements, but all those previously made by reliable observers. They provided, in compiled and critically reviewed form, the observational data on which orbital studies were based.

Aitken also wrote the standard book on his subject, *The Binary Stars*, discussing them from every point of view and summarizing all available knowledge on them. He published the first edition of this book in 1918, and revised and updated it for a second edition in 1935. Two generations of astronomy students learned their double stars from these books.

Aitken was an excellent double-star observer, but his methods were highly routine and hardly changed in forty years. There were no photographic plates to develop or measure, very little analysis or reduction, and essentially no thinking at all. He had plenty of time for other astronomical activities. One of his great loves was the Astronomical Society of the Pacific, the combined amateur and professional organization that Edward S. Holden had founded in 1889. Aitken joined it in 1894 while he was still a faculty member at the University of the Pacific, and remained a member until his death fifty-seven years later. Nearly all of the senior Lick Observatory astronomers of the early days were presidents of the society at one time or another. Aitken served two terms, one in 1898 and another in 1915. He was six times the society's vice president, as well as being its secretary for thirteen years, a director for forty-six years, and a member of the publication committee for fifty years. His Williams College training and early experience as a teacher made him a prolific writer and lecturer to amateur astronomy groups, schools, and all kinds of societies.

As the outstanding authority in the world on double stars, Aitken received many honors, not only the Bruce Medal of the Astronomical Society of the Pacific but also the Lalande Medal of the French Academy of Science and the Gold Medal of the Royal Astronomical Society of London.

Although the main business of Lick Observatory was always astronomical research, Aitken never forgot that to the public it provided a fascinating glimpse into the heavens. The observatory was open to visitors every Saturday evening and, weather permitting, anyone who appeared then got to look through the 36-inch refractor. To many it was an unforgettable experience. Aitken himself, or one of the other astronomers, would give a little talk about the wonders of the universe to while away the time for those waiting in line for a quick view through the big telescope.

Distinguished visitors received individual attention; they were welcome whenever they could come, and were given special tours. Several royal personages, beginning in Floyd and Fraser's days with the King of Hawaii, made sight-seeing visits to Mount Hamilton, as did many more famous politicians. Surely the most appreciative and satisfying comments were those of Winston Churchill, who came to Lick Observatory in 1929. Then fifty-four years old, he had just been turned out of his office of Chancellor of the Exchequer, along with the rest of the Conservative government headed by Prime Minister Stanley Baldwin. Churchill, accompanied by his brother Jack, his son Randolph, and his nephew John, was on a lecturing and writing tour of Canada and the United States.

On September 11, 1929, they drove up Mount Hamilton with William H. Crocker, their host in San Francisco, for a visit to the observatory. The party had dinner with Campbell, who had come up for the occasion, Aitken, and other senior staff members. Then they all went to the telescope. It was a beautiful fall evening, with a steady atmosphere and excellent seeing, permitting the use of high magnifying power at the 36-inch refractor. Churchill was amazed by what he saw through the telescope. He wrote his wife, in England:

> First of all they showed us the planet Saturn. To the naked eye this looks like any other star but when, in the great dark hall, I put my eye to the telescope—an object of sublime beauty was disclosed. I enclose a picture postcard of what I saw but you must imagine this all glowing like a brilliant lamp. This spectacle took my breath away. Although I had often heard of the ring of Saturn, I had no conception of the perfectness and splendor of this orb. Indeed, I thought at first that it was the reflection of a powerful electric light which they had forgotten to turn out and could not realize that I was looking at a world 800,000,000 miles away.

Next they looked at the moon, an even more spectacular sight, especially since it was at first quarter and through the 36-inch they could see the startling contrast of the shadows of the mountains along the terminator. "Being little more than one quarter million of miles away, one felt one could almost touch her," he wrote. Evidently they got a little lecture on spiral galaxies, for Churchill ended on a reflective note:

> It appears that there are several million universes, each consisting of hundreds of millions of suns equipped with planets which again are attended by moons. After contemplating the heavens for some hours one wonders why one worries about the Epping Division [the district he represented in Parliament].

The next day they drove on to Pebble Beach, and the day after that to San Simeon, where Churchill contemplated Marion Davies, William Randoph Hearst's mistress, with considerably less awe.

As associate director of Lick Observatory for seven years Aitken was responsible for the day-to-day operations, but submitted all major decisions to Campbell as recommendations, for his approval. The president retained a very deep interest in astronomy, and although he had no time for research himself, he did not hesitate to express firm opinions to Aitken as to what type of work others should and should not do. Throughout his entire term as president, Campbell retained not only the title of director, but also the reality of possession of the director's house on Mount Hamilton. He had lived in the house, which was first built for him as a young staff member, since 1893. At the height of his success as director, according to persistent rumors on Mount Hamilton, he had asked the regents to build a larger house for him, with more rooms for his servants. When the regents declined, pleading the necessity of using the funds for more urgent projects, Campbell had the observatory maintenance employees, who worked under his direction, "improve and enlarge" the house. This essentially meant building a new house around the existing one, with teakwood floors in the living and dining rooms. Upon Campbell's return from the Wallal eclipse expedition in 1922 he found that the observatory carpenter had been repairing his own house, instead of working on enlarging the director's as ordered. For this Campbell fired the man on the spot. The house, when completed, was appraised at over twice the value of the next most expensive house on the mountain, and over three times the average value of the houses in which the other staff astronomers lived.

In the director's house Wallace and Elizabeth Campbell had lived, raised their children, and entertained countless guests, many of them rich friends of the observatory and the university, and famous personalities as well. Campbell kept the house when he went to Berkeley, intending to return more often than he did, but using it occasionally as a weekend retreat or for a brief vacation. Elizabeth Campbell would come more frequently, usually finding what to her by then critical eye seemed some failure or breakdown in the cleanliness or furnishings of the house. She would report it to her husband, who would in turn direct Aitken to have the fault corrected. The Campbells maintained that the Aitkens did not want or need the director's house (though rumors flourished in the little mountain community that Jessie Aitken wanted it very much indeed). Thus even when Campbell retired as president, the regents, no doubt at his hint, let him retain the house "in recognition of his long and excellent service."

Hence even as director in his own right, Aitken appeared as more or less a caretaker figure. He was more restrained than Campbell; less forceful and dominating. By 1930 the Great Depression had begun. Money became a constant problem for the university and especially for the observatory, which had practically no students and was far from the seats of the mighty in Berkeley. Aitken could hardly be blamed for not building up Lick Observatory; he had to devote all of his energies to keeping it alive and operating on a minimal budget.

Sometimes, in spite of his best efforts, the financial realities hurt Lick Observatory badly. One case in point is that of Gerrit (later Gerard) P. Kuiper, a brilliant young Dutch astronomer who came to Mount Hamilton in 1933 as Aitken's protege and hand-picked successor. Kuiper, as an eager young student in Leiden, had first written to the world-famous double-star observer in 1929. Just 24 years old, Kuiper was determined to make double stars his own life work. He had begun measuring them himself with the 10½-inch Leiden refractor just three months previously, and he sent his first results to Aitken for his criticisms and advice. Already then Kuiper had formulated his plan to make a "complete" survey to find how many stars in a particular volume of space down to a very faint absolute magnitude (or luminosity) are doubles, and of these what proportions of them have various differences in brightness between their two components, various orbital sizes, and so on.

Aitken found the young student's work excellent. He cautioned him not to push his measurements to pairs too close for his small telescope under the poor Dutch skies, but encouraged him to go on with his research. Kuiper replied gratefully, accepting all of the director's advice, saying he would follow it carefully in the future, and explaining that he had only tried the measurements of the close pairs to test himself under the most difficult conditions, after he had practiced long with the easier, wider pairs. Kuiper's professors in Leiden, Ejnar Hertzsprung and Willem de Sitter, recommended their young student to Aitken as an outstanding prospect. In January 1932, even before Kuiper had completed the requirements for his Ph.D. degree in Holland, Aitken offered him a fellowship to come to Lick and carry on his double-star research with the 12-inch and 36-inch refractors.

Kuiper preferred to wait until he had finished his thesis, but indicated that he wanted very much to work at Lick and would be ready the following year. That spring Aitken traveled to England, to receive the Gold Medal from the Royal Astronomical Society of London. He took advantage of the opportunity to visit de Sitter and Hertzsprung in Leiden, becoming even more impressed with their energetic, thorough student. In 1933 Aitken again offered Kuiper the Martin Kellogg Fellowship. It was one of two that the Lick director controlled, based on an endowment left to the observatory by the former president of the University of California. The other was the Alexander Morrison Fellowship, endowed by the will of a wealthy San Franciscan whom Campbell had cultivated and awakened to the needs of astronomy. The income from these two funds enabled the Lick director to bring promising young scientists to Mount Hamilton as postdoctoral fellows, independently of state funds. The previous year the Morrison Fellowship had been available, but when Kuiper declined it, Aitken had given it to another prospect, who would keep it for two years, the usual period. Thus in 1933 only the Kellogg Fellowship was available. Its income was low, because of the depression, and Aitken could offer Kuiper only $950 for the academic year. The young Dutch astronomer got an additional $200 grant in Holland, and the combined sum was enough to get him to America and keep him alive in the dormitory on Mount Hamilton for eleven months.

Kuiper completed his degree requirements in Leiden on June 30. He had decided his future would be in America, and he brought with him his books, manuscripts, mathematical and astronomical tables, and calculations in progress. He shipped separately the set of objective gratings that he had used for his double-star work at Leiden. Each grating was an arrangement of thin, parallel, black rods, mounted so it could be slipped as a unit over the front surface of the lens ("objective") of a telescope. It produced by diffraction a series of images of each star, not only a central bright one but also several fainter ones on either side of it. The difference in magnitudes between successive images depended only on the spacing and thickness of the rods, and could be calculated accurately. Thus the grating allowed Kuiper to measure the magnitude difference between two stars of very different brightness, such as the two components of a double star, by comparing a secondary image of the brighter star with the primary image of the fainter one.

Kuiper planned his trip with typical meticulous thoroughness. He informed Aitken that he would arrive at Oakland on July 31, at 7:51 A.M. and that "apart from accident" he would get to the observatory before noon the next day (when his fellowship was to begin) by the regular "stage." There were no accidents and Kuiper was there at the appointed time.

The young "postdoc" brought to Lick the careful Dutch statistical approach, going back to J. C. Kapteyn, in which he had been trained by Hertzsprung and de Sitter. His program was based on a well-defined "sample," and all his methods were framed to eliminate observational biases and to get accurate statistical results on the types of stars that exist in the universe. Furthermore, unlike Aitken, Kuiper was not content simply to measure the separation and position angle of each double-star pair. He wanted to find the nature of the stars. In addition, therefore, he measured the magnitudes of the individual components (using his gratings if necessary) and obtained spectrograms as well. The separations and magnitudes he determined with either the 12-inch refractor (for bright and wide pairs), or the 36-inch if it were necessary, while he took the spectrograms at the Crossley reflector with a small slitless spectrograph, an improved version of the instrument Keeler had conceived thirty-five years before.

Kuiper worked very hard. For his second year Aitken was able to give him the Morrison Fellowship, which had by then become vacant, providing a little more money. The energetic young postdoc found that a very high percentage of the most common stars in space, the low-luminosity ones that are less massive than the sun, are double. An appreciable fraction of the fainter members of the pairs, he learned, were "white dwarfs," stars of a type that had only first been recognized and interpreted theoretically a few years before, with incredibly high densities and very small diameters.

Two years was the limit anyone was allowed to hold a fellowship at Lick. These appointments were meant as opportunities for young astronomers to come to Mount Hamilton to work, broaden their horizons, and transfer some of their fresh ideas to the permanent Lick staff. The fellowships also gave the Mount Hamilton astronomers a chance to look over the new Ph.D.s from all over the world and pick out the best as the staff members of tomorrow. Aitken was to retire in the summer of 1935. He and the other senior Lick astronomers chose Kuiper to be his successor. The young Dutch physical double-star observer would take the place of the older American orbital double-star observer. Kuiper was eager to stay as a permanent faculty member and Aitken recommended his appointment to President Robert G. Sproul in Berkeley. However, the depression prevented it from being made. Tax collections were very low in California in 1935, and the legislature was threatening to reduce the appropriation for the university. New appointments suffered, particularly in purely research positions. Sproul told Aitken that Kuiper could not be appointed.

Through his Dutch connections, Kuiper was offered a position at the Bosscha Observatory in Java. Though he wanted to stay in America, he had no other choice and he very nearly took this job. However, at the last moment he received an offer of a temporary appointment at Harvard, which he accepted. His last two months at Mount Hamilton, before the fall semester began at Cambridge, he spent completing his observations as a research assistant paid from the William H. Crocker Fund, the third and smallest source of money the Lick director could dispense.

At Harvard, with no large telescope under a clear, stable sky for double-star measurements, but with numerous students to teach in an elementary astronomy class, Kuiper thought often of Lick Observatory. But he did his work enthusiastically and wrote up the results of his fellowship research. Within a year he was offered a position to observe at McDonald Observatory, then under construction by the University of Texas. The new observatory when completed was to be operated for many years by the University of Chicago. Kuiper moved to Yerkes Observatory and continued his double-star spectral classification and his investigations of low-luminosity stars there and at McDonald. After World War II he and Harold C. Urey at the University of Chicago led a reawakening of interest in the physical study of the origin of the solar system and the planets. This subject, dealing as it does with the division of angular momentum between a star (the sun) and its planets, is very closely related to Kuiper's early study of double stars, and how their angular momentum is divided between them. Kuiper continued his research on the solar system at the Lunar and Planetary Laboratory, which he founded at the University of Arizona, after leaving Yerkes in 1961.

Another European astronomer who came to Lick Observatory before Kuiper, and stayed longer, was Robert J. Trumpler. Born in Switzerland and trained in Göttingen, he had come to the United States in 1915, and to Lick Observatory (as a Martin Kellogg Fellow) in 1919. He worked as Campbell's assistant at the Wallal eclipse. Afterward Trumpler measured and reduced the plates they had taken there, which confirmed so beautifully Einstein's general theory of relativity. But Trumpler's main interest was in the study of star clusters, not in relativity.

He had worked at Allegheny Observatory with Frank Schlesinger, the master of modern positional astronomy. Trumpler had made his special project at Allegheny the determination of the "proper motions" (or motions across the plane of the sky) of the stars in the Pleiades, the nearby cluster whose brightest members can be seen in the autumn sky as the "Seven Sisters." All the stars in a cluster clearly must have nearly the same velocity in space or the grouping would long ago have dispersed; measuring the proper motions of all the stars in the region allowed Trumpler to weed out the foreground and background stars that do not really belong to the Pleiades.

At Lick he studied many clusters. In each one he obtained spectra of as many of the stars that belonged to it as he could. Using the new methods of photographic photometry, Trumpler also measured accurately the magnitudes of all these stars. Each individual cluster is obviously relatively small; all its stars are at very nearly the same distance from us. The fainter stars in a cluster are not fainter because they are further away from us than the brighter stars; they are fainter because they are intrinsically less luminous. Trumpler could therefore plot the "Hertzsprung-Russell diagram," the relationship between spectral type and intrinsic luminosity, or "absolute magnitude" as astronomers call it, for the stars in each cluster, strictly from observational data.

This diagram is one of the great tools of modern astronomy. As Hertzsprung and Henry Norris Russell had discovered more or less independently not long before, the spectral type of a star is correlated closely with its absolute magnitude. Thus by determining the spectral type of a star (for instance that it has a spectrum like the sun's, called G2 dwarf), the astronomer immediately knows its intrinsic luminosity (that is, that its absolute magnitude is + 4.6, like the sun's). Comparing this absolute magnitude with the measured ("apparent") magnitude of a star tells how far away it is, for the farther it is the less light we get from it and the fainter it looks.

Trumpler applied this reasoning to the spectral types he had determined and the magnitudes he had measured in the individual clusters. He found that the stars in each cluster did indeed fit on a Hertzsprung-Russell diagram of the general form already known, thus confirming that the concept was a good one. He could then determine the distance of each cluster from his measurements by the relative faintness of the stars in the cluster at each spectral type in comparison with the standard absolute magnitudes given by the Hertzsprung-Russell diagram.

By 1930 Trumpler had measured the distances of one hundred clusters in this way. The clusters appeared to be more or less similar to one another. However, their diameters, which he could determine from their angular sizes in the sky and their distances, did not seem to be similar. According to the data reduced in this way, the more distant clusters were larger than those closest to the sun. There seemed to be a smooth increase of cluster diameter with apparent distance from the sun.

Trumpler realized that this apparent correlation could not in fact be true. The sun was not at the center of our galaxy or in any special position in it, and there was absolutely no reason why clusters' diameters should depend upon how far away from the sun they are. Some other effect had to be coming into play. Trumpler recognized that it must be the absorption of light by solid particles in space between us and the distant clusters. Such absorption would make distant objects appear fainter than they would be if space were absolutely transparent. The apparent magnitudes of the stars in a cluster depend, Trumpler saw, not only on how far away the cluster is, but also on how much absorbing matter there is between us and it. The greater the amount of absorption, the fainter the stars appear, and the farther away the cluster appears to be if we do not recognize that absorption is present.

Thus Trumpler proved that there is absorbing material in our galaxy between the stars, "interstellar matter" in modern terminology. Furthermore, he showed that the density of interstellar matter is more or less uniform, for the more distant clusters showed more absorption then the nearby ones, approximately in proportion to their distance. By making this reasoning quantitative, Trumpler was able to determine the average amount of interstellar absorption in our galaxy.

As he recognized, the clusters he measured, called galactic clusters by astronomers, lie close to the Milky Way in the sky; that is, they are located in the plane of our galaxy. The absorption Trumpler measured thus refers to interstellar matter in the plane of the galaxy. The much more distant "globular" clusters are mostly far from the plane of the galaxy and, as Trumpler said, they therefore show these absorption effects to a much lesser extent than the galactic clusters do. In this way Trumpler proved that the interstellar matter is strongly concentrated to the galactic plane. He had found additional evidence reinforcing the analogy between our galaxy and the distant spiral nebulae, first recognized in great numbers by James E. Keeler thirty years before, and studied in detail as "island universes" or galaxies by Heber D. Curtis between 1910 and 1920. Much of our modern understanding of interstellar matter dates from Trumpler's classic paper.

Trumpler and his wife had five children, a sizeable portion of the student body of the one-room school in the little mountain community. The Trumplers kept a flock of chickens and sometimes a goat or two, which caused occasional stress between the normally quiet, retiring Trumpler and the director. Both Campbell and Aitken wanted to keep the livestock out of sight and out of earshot. Most of the time conditions were tranquil, however. The observatory families had an active social life on the mountain, centering around the school, visits, and informal parties. However, as the Trumplers' children grew to high school and college age, the parents, as many before them and since, decided it was time to move to a larger community. Trumpler bought a house in Berkeley, and arranged to live there with his family each school year, beginning in 1935. He did some teaching, but spent most of his time on research, making periodic trips to Mount Hamilton to observe with the telescopes. In 1938 Trumpler transferred permanently from the Lick Observatory staff to the Berkeley faculty, and he spent the rest of his career on campus. Although his "preliminary results" on the galactic clusters had such an important role in pushing out our understanding of our galaxy as a representative spiral and of the absorbing interstellar material within it, he never finished and published the monumental "complete discussion" that he had planned. Trumpler retired in 1951 and died in 1956.

Besides Kuiper and Trumpler, another very important research worker who was at Mount Hamilton during the Aitken years was Donald H. Menzel. He was the first theoretical astrophysicist on the Lick staff. All of the early Lick astronomers were basically observers, a few of them (such as Keeler and Henry Crew) trained in laboratory physics, but mostly in observational astronomy and mathematics (like Campbell, Aitken, and Curtis). Trumpler had a very high-class scientific education in Göttingen, but even he had concentrated mostly on astronomy and mathematics, and had no pretensions toward developing new theoretical ideas.

By the 1920s, however, it was becoming clear that theorists who understood and could use modern physics would be able to make important contributions to astronomy. The outstanding example was Henry Norris Russell, the Princeton professor who had independently developed the ideas of the Hertzsprung-Russell diagram. Fantastically brilliant (to this day he is the only person who ever received his Princeton undergraduate degree "insigne cum laude"), he had done important work not only in statistical astronomy but in such physical subjects as the analysis of double stars, stellar atmospheres, the abundances of the elements, and the internal structure of the stars. He was even a pioneer in the application of quantum theory to atomic spectroscopy.

Russell was no ivory-tower theorist. He wanted to get his hands right on the latest observational data. He made frequent visits to Harvard, where his former student Harlow Shapley was director of the observatory, and nearly every summer made a trip west to work at Lick, Mount Wilson Observatory, or Lowell Observatory, in Flagstaff, Arizona. Russell was full of creative new ideas, and always galvanized the Lick observers into new channels of thought.

In 1926, when Aitken had an opening for a new faculty member, he and Campbell decided to fill it with a trained astrophysicist. The primary impetus was to get someone to reduce, analyze, and discuss the spectrograms of the chromosphere and corona of the sun that Campbell had been accumulating at eclipses since 1898. Another area in which some new physical ideas were clearly needed was William H. Wright's program on planetary nebulae. Aitken wrote Russell and asked him if he could recommend anyone "who might perhaps be available as an assistant astronomer at the Lick Observatory, with duties in astrophysical lines. I am looking for such a man and have not found him."

Russell immediately recommended Menzel as "the best chap that I can think of. He is really decidedly good, and knows quite a bit of astrophysics." Not surprisingly Menzel had been Russell's student—for in the 1920s there was essentially no other place in the United States but Princeton where a person could study astrophysics. A native of

Colorado, Menzel had studied astronomy, physics, and mathematics at the University of Denver, where he had earned his bachelor's degree in 1920 and an A.M. in 1921. In his senior year and as a graduate student he had served as a part-time instructor of mathematics at the university, and had also done some substitute teaching in the Denver high schools.

Then Menzel had gone to Princeton, where he had studied and worked with Russell for three years. There he was a graduate student in the very early days of quantum theory, as the Bohr semi-classical picture of the atom gave way to the Schroedinger wave equation and the Heisenberg uncertainty principle. The whole world of astrophysics opened out before Menzel. Understanding almost everything about the stars seemed possible. Russell, an extremely quick learner and brilliant puzzle solver, was making major contributions in spectroscopy and its applications to quantitative understanding of the stars. Menzel, a bright, quick student, learned atomic theory, ionization calculations and all the other advanced astrophysical tools that Russell could teach him. Each summer Russell sent Menzel to the Harvard College Observatory to apply his new knowledge to actual problems on the spectra of stars. Menzel's thesis, done under Russell's supervision, was on the interpretation of the spectral classification of the stars and the physical meaning of the Hertzsprung-Russell diagram. At the very same time Cecilia Payne (later Payne-Gaposchkin) was working along similar lines as a Harvard graduate student; she concentrated on the hot stars and Menzel on the cool stars.

When Menzel completed his Ph.D. in 1924, he got a job at the University of Iowa, and after one year there found a better position at Ohio State University. Both were teaching jobs with little time for research at universities with essentially no observational facilities. Thus Menzel was eager to move on to Lick when Aitken offered him a position in 1926. A complication was that Menzel was getting married that summer, and he and his bride planned an extended honeymoon and a leisurely trip west. Menzel asked if he could not wait to start work until the fall. Aitken insisted that although the job could begin one month later than the normal July 1 starting date, and Menzel could postpone his arrival one week beyond August 1, he absolutely must be on the job by Monday August 9. Menzel was there, but he must have wondered just why it was so necessary for him to start working and drawing his pay in August rather than October. It was only the first of many differences of opinion he had with the associate director.

Menzel's main work was on Campbell's eclipse plates. The young astrophysicist did an excellent job with this great collection of observational material, and his research led to the beginnings of a real understanding of the physical nature of the chromosphere. In addition,

Menzel was very interested in the problems of planetary nebulae. He wrote an excellent paper, reviewing all the various ideas for interpreting them physically that were then current. In it Menzel showed an excellent knowledge of observational data and of theoretical methods for analyzing them. One of his main points was that, contrary to the opinion of most astronomers at the time, the central stars of planetary nebulae, though very hot, must be much less luminous than the normal "main-sequence" stars with similar temperatures. Menzel connected the planetary-nebula stars with the recently discovered "white-dwarf" stars, a daring extrapolation that today is one of the central tenets of our knowledge of stellar evolution. Menzel's paper on the planetary nebulae was very physical, very forward looking, and very nearly completely correct.

However, not much came of it at Lick. Wright, though apparently interested in Menzel's ideas and willing to discuss them with him, did not offer to share his data or collaborate with the younger man. Wright did not modify his observing program to attempt to follow up any of the Princeton theoretician's ideas. Menzel could not start a competing program of his own on planetary nebulae. It was only after he left Lick, several years later, that he began his epochal series of theoretical papers on the physical conditions in planetary nebulae.

On Mount Hamilton Menzel was a beehive of activity, reading, studying, and offering suggestions. At Campbell's request he reviewed carefully the long series of measurements that Charles G. Abbot had made of the total flux of radiation from the sun. Abbot claimed to have detected solar variations that were correlated with the number of sunspots, which vary in time over the "sunspot cycle" of approximately eleven years. Menzel discussed the measurements and the theory behind them with equal lucidity. Although he thought there might be some evidence for variation, he was highly skeptical of the purported observational correlation with the sunspot cycle. Like most of Menzel's work at Lick, it was very good, and strongly based on correct physics.

By the summer of 1927, less than a year after his arrival, Menzel was getting restless on Mount Hamilton. He arranged a period of leave to go to Mount Wilson Observatory in southern California to collect observational data on the identifications of solar absorption lines for use in his own research on the spectrum of the chromosphere. Immediately after his return he went to Reno for the summer meeting of the Astronomical Society of the Pacific. The next year he arranged to trade places for one semester with R. Tracy Crawford of the Berkeley astronomy department. Menzel taught on the campus, while Crawford did research on Mount Hamilton. This visit solidified Menzel's contacts with the Berkeley physicists, especially the noted spectroscopist Raymond T.

Birge, and the theoretician J. Robert Oppenheimer. During his semester in Berkeley Menzel also prepared, at Aitken's request, lists of physics books that should be bought for the Lick Observatory library. Until then its collection had been concentrated almost entirely in astronomy.

That summer Menzel went to Mount Wilson Observatory again for three weeks, followed by a trip east that included a vacation in Colorado with his wife and baby daughter, a visit to Harvard where he gave a talk on his research, and the meeting of the American Astronomical Society in Amherst, Massachusetts, where he presented two papers. Along the way he stopped in New York to visit several motion-picture producers, whom he was trying to interest in making an astronomical film.

Aitken was accustomed to astronomers who were happy to remain on Mount Hamilton all their lives, observing three or four half-nights per week, and reducing their data. He did not approve of the young theoretician's frequent travels. All kinds of strange things were happening at the once quiet observatory. A newspaper feature agency in New York addressed a letter to the "Director of Publicity" of Lick Observatory, asking for a photograph of Menzel to publish with an article about his work on the temperature of the moon. The very thought that Lick Observatory might have a person whose job was in publicity was repugnant to Aitken.

On April 28, 1930, there was to be a brief solar eclipse in California and other western states. The moon was so far from the earth, and its angular size consequently reduced, that the eclipse would be only annular along much of the path; that is, the moon would not completely cover the sun but would leave a bright ring all around the edge, even at the time of maximum coverage. But for about a thousand miles, from central California to southwestern Montana, the eclipse would be barely total. It would last only a few seconds, so long-exposure direct photographs or spectrograms of the faint corona were out of the question. However, totality would last long enough to obtain spectrograms of the brighter chromosphere, and Menzel wished to do so, to get well-calibrated data for comparison with Campbell's earlier uncalibrated plates.

Aitken sent a group including Joseph H. Moore, Menzel, and C. Donald Shane to observe the eclipse at Camptonville, in the Sierra northeast of Sacramento. Moore, a senior staff member, was in charge. Soon after they arrived at the site, Menzel sent a message back to Aitken, requesting his permission to write a story for the *San Francisco Chronicle* on the Lick party's plans. Aitken gave his approval grudgingly, but reminded Menzel not to make any claims for vast discoveries before they had been made, and told him "to let Dr. Moore read over the article and suggest any necessary changes" before releasing it. They had rain

off and on for days, right up to the night before the eclipse, but the
Lick good luck held up and the skies cleared before the big event the
next morning. They got their eclipse spectrograms.

Menzel worked on many other topics at Lick. He began to apply the
ideas of deviations from thermodynamic equilibrium, so important in
the corona and in planetary nebulae, to understanding Wolf-Rayet stars.
He persuaded Moore to collaborate with him in measuring the rota-
tional velocity of the planet Neptune by the same spectroscopic method
Keeler had applied to the rings of Saturn a generation earlier. But in
the summer of 1932, after many more trips away from Mount Hamilton, Menzel received an offer of a faculty position at Harvard. His
interests were all strongly directed toward theoretical research, a university campus, graduate students, and teaching, and he welcomed the
offer. Birge and Armin O. Leuschner, the chairman of the Berkeley astronomy department, urged Aitken to promote Menzel to associate astronomer (the equivalent of associate professor) if that would keep him
at Lick. Probably Menzel would have gone in any case, but Aitken
would have none of it. He recognized Menzel's outstanding abilities
and potentialities, but he refused to promote him over the heads of his
older colleagues. Aitken told President Sproul that Menzel was good,
but that he was "not so much interested in actual observing as in theoretical work." Aitken felt "that it is of primary importance that the
men connected with Lick should be first and foremost observers." Menzel left for Harvard soon afterward.

There was another eclipse of the sun, this one with a fairly long
totality, in the eastern United States on August 31, 1932. A Lick Observatory group, consisting of Moore, Wright, Menzel, and Shane went
to Fryeburg, Maine to observe it. It was the first long total eclipse since
1893 that Lick astronomers observed at which they did not use the
Schaeberle camera. Wright had concluded that with the fine-grain photographic emulsions by then available, shorter focal-length cameras
were sufficient to show all the details that could possibly be recorded.
Moore and Menzel obtained spectrograms of the corona, and after the
eclipse, when the others returned to Mount Hamilton, the new Harvard
astronomer remained in Cambridge. He had a long and successful career
there, developing many important new theoretical ideas about the solar
chromosphere and gaseous nebulae. When Shapley retired in 1952,
Menzel succeeded him as director of Harvard College Observatory, and
remained in the post until 1966. He retired himself in 1971 and died
in 1976.

Another scientist who made a notable contribution during his short
stay at Lick Observatory was Nicholas T. Bobrovnikoff. Born in Russia,
he fought in his country's army as a young officer in World War I, and

then in General Anton Denikin's White Army against the Bolsheviks. Bobrovnikoff was severely wounded, suffered frozen feet, and nearly died of typhus before the White Army collapsed in March 1920. He managed to get out of Russia on one of the last hospital ships just before the final defeat. Against great odds Bobrovnikoff made his way to Czechoslovakia, where he succeeded in getting an undergraduate degree in astronomy at the Charles University in Prague in 1924. By another great stroke of fortune he obtained a fellowship for graduate work at Yerkes Observatory of the University of Chicago. His Ph.D. thesis was an astrophysical study of the nature of comets, based on the large backlog of cometary spectrograms that had been accumulated at Yerkes, going back to 1908. The most complete series included fifty-three spectra of Halley's Comet, taken in 1910 when it had last passed close by the sun.

When Bobrovnikoff completed his Chicago thesis in 1927, he was appointed to the Martin Kellogg Fellowship at Lick. Edwin B. Frost, the Yerkes director, Otto Struve, its rising young observational astrophysicist, and George Van Biesbroeck, its senior double-star observer, had all recommended him to Aitken. Bobrovnikoff wanted to go to Mount Hamilton, because he knew there was an unexcelled collection of direct photographs of Comet Halley crying for attention there. Curtis, with his assistant Charles P. Olivier, had taken over two hundred direct plates of the comet with the Crossley reflector in 1909–1910, and almost as many more plates with various smaller telescopes. There were also many plates from the Lick southern observatory in Chile. In addition there were numerous spectrograms that had been taken by Wright. Bobrovnikoff supplemented this material by borrowing additional spectrograms and direct plates that had been taken at Mount Wilson Observatory by Walter S. Adams and George W. Ritchey. Many astronomers had worked countless hours to secure all this observational material on one famous comet, but none of them had ever taken the time to analyze and discuss it all thoroughly. This is the task Bobrovnikoff set himself.

Measuring the direct photographs, he could find individual features in the tail of the comet and follow their motion from day to day. From these measurements he could compute the velocities and then the accelerations of the particles in the tail, and thus the repulsive force from the sun that was acting upon them. Bobrovnikoff found a wide variety of values of the repulsive force. They could not all be due to light pressure, as many astronomers had assumed up to that time. He discovered that the ratios of the strength of the repulsive force to the strength of the solar gravitational force cluster around two values, one much larger than the other. Today we know the larger values result from the interaction of the tail with the solar wind, the smaller, from light pressure on dust particles in the tail.

Bobrovnikoff emphasized that the larger repulsive force acted on the comet's long straight tail, and that it was composed of fine "streamers" (filaments) and condensations, all of which show the spectrum of the molecule CO^+. He described the "striking analogy" between the motions of matter in the long straight tail and in solar prominences. All of these were observational clues to the importance of charged particles in the solar wind interacting with cometary material, as we realize today but Bobrovnikoff could not.

He also described clearly the fact that the spectrum of the molecule CN in Halley and other comets arises in very low-temperature material, while the C_2 spectrum does not show this effect. This is understood today to result from the operation of quantum mechanical selection rules, which depend on whether a molecule is composed of two similar or different nuclei.

Bobrovnikoff spent two years at Lick Observatory, reducing and analyzing all the material on Comet Halley. At the same time he carried out several smaller projects, all concerned with the physical properties of comets, asteroids, and planets. Then he obtained a National Research Council Postdoctoral Fellowship, on the recommendation of Aitken, for a final year at Berkeley. There he completed the analysis of his measurements and wrote up his long paper on "Halley's Comet in its Apparition of 1909–1910." It was a landmark contribution that pointed the way for the modern astrophysical study of comets. It appeared as the second part of *Lick Observatory Publications,* volume 18; Menzel's equally long paper on the chromosphere was the first part of the same volume.

In the summer of 1930 Bobrovnikoff married Mildred Sharrer, who had been the teacher in the one-room school on Mount Hamilton while he was at the observatory. He got a job as an assistant professor at Perkins Observatory of Ohio Wesleyan University, where a 69-inch reflecting telescope was under construction. Ultimately Bobrovnikoff became director of Perkins, and played a major role in bringing about an agreement by which Ohio State University eventually took over its operation.

Two of the most astrophysically-oriented theses done at Lick Observatory in the Aitken years were those of Louis Berman and of Fred L. Whipple. Berman had earned his B.S. and M.S. degrees at the University of Minnesota before coming to California in 1927. He was trained in the orbital and positional astronomy that Leuschner and Aitken loved; they respected him even more because he plunged into double-star observing, and demonstrated his abilities to make emergency repairs to a micrometer eyepiece when necessary, and to get reliable results at all times. But Berman's heart was in astrophysics. He did his thesis on the physical interpretation of planetary nebulae, basing his study partly on spectrograms he took himself with the Crossley reflector, and partly on plates previously taken by Wright. Berman obtained monochromatic images of each nebula in the light of individual spectral lines. Summing up all the light in these images, he could measure quantitatively the total emission of the nebula in the lines of individual ions. The interpretation of these measurements in physical terms gave important insights on the structure, temperature, and ionization processes in the nebulae. Berman's thesis topic had been suggested to him by Menzel, who guided his work. Berman's paper was a very important early reconnaissance of the physical nature of gaseous nebulae, from the observational point of view.

After receiving his Ph.D. in 1929, Berman returned to Carleton College in his native Minnesota as a faculty member. After a few years he came back to California, to a position at San Mateo Junior College, not far from Lick Observatory. The Great Depression was in full swing, and Berman nearly lost his job in San Mateo in a severe economy measure at the college. He managed to hang on, and even to do further important research on planetary nebulae and on the carbon-rich variable star R Coronae Borealis during summers at Lick Observatory. Berman taught at San Mateo, and then at San Francisco (first Junior College, now State University) for many years, and he wrote a widely used textbook in astronomy with a strong astrophysical emphasis.

Whipple had done his undergraduate work at UCLA. Like Berman, he was an expert in the orbit computation that was the true test of astronomical competence in Leuschner's eyes. Whipple and Ernest C. Bower, another graduate student at Berkeley, made one of the first calculations of Pluto's orbit in 1930, soon after the ninth planet had been discovered and announced at Lowell Observatory. Like Berman, Whipple did his thesis under Menzel's guidance. It was an astrophysical study of two cepheid variables. Whipple started from the known fact that they were pulsating stars. His spectroscopic observations, taken with the 36-inch refractor, were designed to measure their color, temperature, and velocity changes simultaneously. By combining these with their measured brightness changes, Whipple could analyze their structure and nature. Many of the methods and theoretical ideas that he used came from Menzel's earlier study of the solar chromosphere. Whipple's thesis was an excellent one. After completing it, he left for a job at Harvard, where he worked on the astrophysics of stars, planets, meteors, and especially comets. He invented the "icy conglomerate" model of comets, and became director of the Smithsonian Astrophysical Observatory.

So Aitken, as associate director and then director, presided over twelve years of Lick Observatory's existence. They were not years of growth. Astrophysics came increasingly to the fore, though traditional astronomy remained important. Some of the most forward-looking research was done by young postdoctoral fellows or short-term staff members, many of them from abroad. Through it all Aitken continued to observe visual double stars. On June 30, 1935, the last day of his service as an active faculty member before his retirement, he observed with the 36-inch refractor until midnight, then closed his observing book and went home to bed. Aiken and his wife left Mount Hamilton the next day and moved to Berkeley, where he remained very active in the Astronomical Society of the Pacific for many years. He was president of the American Astronomical Society from 1937 until 1940, and he died in 1952 at the age of eighty-seven.

12

The Last of the First

1935–1945

William H. Wright was the quintessential Lick Observatory director. Tall and dignified, quiet and deliberate, he was a master of observational spectroscopy. He idolized James E. Keeler, and poked gentle fun at W. W. Campbell's imperious ways and tightness with a dollar. Wright tackled hard research problems, and stuck to them for years. He had firm ideas on what to do and how to do it. He went his own way, with very little theoretical guidance or even discussions with other scientists. And he got important results.

Wright was born in San Francisco in 1871, the descendant of two distinguished old southern families. He studied civil engineering at the University of California, graduating with a B.S. in 1893. But he had become interested in basic science, and he stayed on at Berkeley as a graduate student of mathematics and astronomy. Armin O. Leuschner, his teacher, was only a few years his senior. Wright spent the summer of 1895 studying and working on Mount Hamilton, and when it ended he was fully converted to astronomy.

In 1896 he got a fellowship at Yerkes Observatory, then still under construction at Williams Bay, Wisconsin. No doubt Wright had hoped to do research with the 40-inch refractor, given to the University of Chicago by tycoon Charles T. Yerkes as "the largest and best [telescope] in the world" that would "lick the Lick," but it was not completed. However, Wright worked with young George Ellery Hale at his Kenwood Observatory in Chicago, and learned the essentials of practical spectroscopy. In his last few months at Yerkes Observatory, Wright helped put the finishing touches on the 40-inch and get it into operation.

Then he returned to California in the summer of 1897, hired by Director Edward S. Holden as a temporary replacement for Campbell, who was preparing to leave for the solar eclipse in India. Wright took over the radial-velocity program while Campbell was away for nearly a year, and Keeler, when he became director, got the young San Franciscan's job converted to a permanent one. Wright's analytical mind, careful attention to detail, and engineering education enabled him to suggest several improvements to the instrumentation that have remained standard features ever since. They included using totally reflecting prisms at the ends of the slit to introduce the standard "comparison" spectrum from an iron arc into the spectrograph, a thermostated heat-regulating system to keep the instrument at constant temperature which would eliminate expansion and contraction and consequent changes in focus, and a method of supporting the spectrograph near its center of mass, instead of at one end, doing away with flexure.

In 1903, after Campbell was injured in testing the 36-inch mirror for the Lick southern-hemisphere observatory, Wright took over for him. He completed the tests, and then headed the expedition to Chile and set up the telescope of the Lick southern station on Cerro San Cristobal in Santiago. His mechanical skills and drive enabled him and Harold K. Palmer to get the telescope and spectrograph into routine operation within only a few months. Wright continued observing in Chile for three years, returning to Mount Hamilton in 1906 with 900 spectrograms of 250 stars. They formed the first installment of Lick data on the velocities of stars in the southern hemisphere.

At Mount Hamilton Wright measured these spectrograms and published the results. He also obtained many spectra of several novae, or "new stars," objects which suddenly flare up to great brightness and then slowly decline to their normal, quiescent state. Their spectra are complex mixtures of absorption and emission lines and bands, showing large and changing velocities of expansion, densities, temperatures, and ionization. Wright's observational results on novae played a major part in helping astronomers understand how these stars throw off shells of gas, which first brighten tremendously, and then fade out as they escape into space.

Following the path blazed by Keeler, Wright also worked for many years on the spectra of gaseous nebulae. With the Lick spectrograph he determined accurate wavelengths for many more of their emission lines than the two brightest ones that Keeler had measured visually in 1890 and 1891. Wright pushed these measurements as far as he could into the ultraviolet (short-wavelength) spectral region, using a spectrograph he had designed with quartz optics, which transmit this radiation much better than glass. He also pushed out into the infrared (long-wavelength) region, using the dyes that were just coming into use in the early decades of the twentieth century for extending the sensitivity of photographic plates. In this nebular work Wright systematically obtained long-slit spectrograms, and also monochromatic images (pictures taken in the light of individual spectral lines) of planetary nebulae. He soon noticed that different images of the same object have different sizes. These results ultimately led to an understanding of how the ionization of the nebular gas decreases with increasing distance from the central star, the source of the high-energy photons that break up the atoms into ions and electrons.

In 1935, when Robert G. Aitken retired as director of Lick Observatoy, Wright was the logical successor. He was appointed to take over the post on the advice of a high-level University of California faculty committee headed by Leuschner. Its other members included several deans, two chemists, a physicist, and a mathematician, all from the Berkeley campus, but not a single member of the Lick staff. Not surprisingly, this committee concluded that Lick Observatory was too isolated, and recommended that the astronomers and most of their activities, except for the actual observing, be shifted to the campus. Campbell, now emeritus director and president of the university, objected strongly to this position from his office in Washington. The committee also recommended that future Lick faculty appointments be made in astrophysics, rather than in conventional astronomy, and that some of the administrative responsibilities of the director be shared with other staff members.

Wright agreed strongly with the idea of adding imaginative young spectroscopists to the staff, and claimed that he agreed with the idea of giving up some of his powers to his colleagues. But the thought of moving the Lick astronomers to Berkeley was anathema to him. He recognized that for many young families Mount Hamilton was not an attractive living place. It was too isolated, especially for the wives and children. As a result some excellent scientists had left the observatory, while others could not be persuaded to accept positions at Lick. But Wright's own proposed remedy was to set up a center of operations in San Jose, much more convenient to Mount Hamilton than Berkeley. He visualized this center as a place where astronomers could measure their spectrograms, analyze their data, and write their papers. Their children could attend schools in San Jose, and they could go up the mountain for a few nights observing each month. In addition, the San Jose center would be the interface between the observatory and the outside world, where supplies could be ordered, delivered, and stored until they were needed on the mountain. Astronomers without children (like Wright himself) and the director in particular would continue to live at Mount

Hamilton. Wright could not recognize any advantage of a Berkeley location for the Lick astronomers.

Wright's plan was tried as a feeble experiment for a short time. One astronomer and a secretary were stationed in San Jose for a year, but it would have been very expensive to build up a third true research center between Mount Hamilton and Berkeley, and the idea soon withered away.

Wright was married but childless, and quiet. His recreations were hunting, riding, and hiking in the California mountains. Basically, he liked people like himself. Fellow astronomer Robert J. Trumpler was Swiss, had a large family of noisy children, and with his wife kept a menagerie of dogs, cats, goats, and chickens on Mount Hamilton. Wright respected Trumpler as a scientist, but was not on close personal terms with him. Almost all their non-astronomical ideas were opposed to each other's. No doubt Trumpler had been planning to move to Berkeley for the sake of his children's education in any case, but it was after Wright became director that he actually made the move. From the fall of 1935 onward, Trumpler lived with his family in a house he bought near the campus, coming to Mount Hamilton for a few days each month to observe, and for the summers when his children were out of school. He was enthusiastic for this arrangement, as he frequently reported; Wright just as frequently, in endorsing his reports, pointed out that Trumpler was actually depending on the bulk of the Lick staff in residence on Mount Hamilton to maintain the telescopes and to observe for him if he could not get there himself.

Wright, and to a lesser degree his predecessor, Aitken, and his eventual successor, Joseph H. Moore, all felt that people whose ancestors had come to America many generations before, as theirs had, were likely to be better astronomers, scientists, and neighbors than more recently arrived immigrants. They subconsciously believed that an English lineage was preferable to a German, Swiss, Dutch, or any other descent, with their attendant "funny sounding" names. Wright, who could trace his ancestry back to early settlers from Great Britain, carried his feelings close to the surface in his role as observatory director.

Thus, Wright was very enthusiastic about Ira S. Bowen, the California Institute of Technology laboratory spectroscopist who could also boast a long American lineage. Bowen, twenty-seven years younger than Wright, had done his graduate work at the University of Chicago, where he was Robert A. Millikan's assistant in his spectroscopy laboratory. When Millikan moved to Pasadena in 1921 as the first chairman of the executive council (effectively president) of the newly founded Caltech, Bowen went with him as an instructor and research assistant. In the new Norman Bridge Laboratory with its superb equipment for ultraviolet spectroscopy, Bowen did the experimental work that led to a whole series of papers by himself and Millikan on the spectral lines emitted by successive stages of ionization of many of the light elements. At just this time, the vector model of atoms and ions was being developed by Henry Norris Russell, Frederick A. Saunders, Wolfgang Pauli, and other physicists, and Bowen used it to analyze and classify the spectra of the individual ions he had measured. He had an unrivalled body of data on the energy levels of these ions. These are the discrete energies an atom or ion can have; a spectral line is emitted when an electron makes a transition from a higher to a lower level, releasing a photon with an energy exactly equal to the difference between the two levels. This energy difference corresponds to a specific wavelength of light, and for this reason each ion emits its own characteristic spectrum.

Spectroscopists realized that they did not observe spectral lines in the laboratory corresponding to every theoretically possible difference between the energy levels of a given ion; instead only particular sets of differences gave rise to actual spectral lines. These were called "permitted" transitions to distinguish them from "forbidden" transitions, which did not occur in laboratory experiments. Bowen was interested in astronomy, and realized that many of the strong spectral lines in gaseous nebulae, including the two strongest lines in the green, measured by Keeler years before, had never been identified with any known atom or ion. Their wavelengths did not agree with those measured in the laboratory from any known source. Some early spectroscopists had thought they might be lines emitted by an ion of an element that did not occur on the earth, and William Huggins had coined the name "nebulium" for this hypothetical element. Exactly the same situation had occurred for the spectral line λ5876, first observed in the sun and later in gaseous nebulae, but not in the laboratory. Its hypothetical source had been dubbed "helium" (sun element). Sir William Ramsay had finally isolated it in 1895 and identified it as a rare gas on the earth. But nebulium had never been found, and by 1927 the spectra of practically all ions had been studied. The physicists' understanding of the periodic table of the elements led to the conclusion that there were no remaining undiscovered elements. Russell had concluded that there was no such thing as nebulium, and that the spectral lines observed in nebulae must in reality be emitted by fairly common elements, but arise in atomic transitions that for some reason could only occur in the very low-density gases that make up these objects.

Pondering these apparent mysteries one night, Bowen realized that there are energy levels in common ions, for instance in O^{++}, or O III as the spectroscopists call the spectrum of twice-ionized oxygen, from which no downward transitions are permitted. In the laboratory, no spectral lines are emitted from these levels. Ions placed in these levels, according to the accepted spectroscopic ideas of the time, could never emit a photon. In the laboratory they would collide with other ions or electrons and give up their energy in this way. But under the extreme low-density conditions in nebulae, collisions would be few and far between. Would an ion stay in one of these excited levels with no permitted downward transitions forever? Perhaps, Bowen reasoned, it might ultimately emit a "forbidden" photon if no collision removed its energy. Struck by the thought in the middle of the night, he got out of bed, dressed, and walked to his laboratory. There he had, in numerical form, all the energy levels derived from his painstaking measurements. Calculating energy differences between the levels, and converting them to wavelengths, he could predict the lines that would be seen if the "forbidden" transitions could in fact occur in the nebulae. Within a few minutes he found that the wavelengths of the two strong green lines, measured so accurately by Keeler in the spectra of planetary nebulae, exactly agreed with the predicted wavelengths of two forbidden lines from an excited energy level of O^{++}. Bowen had discovered that "nebulium" was really doubly ionized oxygen. That same night he was able to identify several other nebular lines, their wavelengths accurately measured and published by Wright, with forbidden transitions in other ions of oxygen, nitrogen, and other fairly abundant elements. It was an outstanding application of laboratory spectroscopy and creative thinking to astrophysics.

Once Bowen had solved the puzzle, he followed it up with further identifications, based in many cases on interpolations or extrapolations of laboratory data. Ultimately, he and other spectroscopists identified hundreds of "forbidden" nebular lines, and once they had been observed, theorists of quantum mechanics were able to extend their calculations and show how these transitions though infrequent are not actually "forbidden."

Bowen showed by clear physical reasoning why these lines are so strong in low-density nebulae, but so weak they are almost impossible to detect in most laboratory sources. Then in 1934 Wright was able to push his measurements of the wavelengths of nebular lines further into the ultraviolet spectral region. This advance was based on aluminizing, rather than silvering, the mirror of the Crossley reflector, which extended its reflectivity to much shorter wavelengths. John Strong had just developed this vacuum aluminization process for telescope mirrors at Caltech. Wright's ultraviolet spectra showed that there are many permitted lines of O^{++} in planetary nebulae, but that their relative strengths are very different from those in any laboratory source. In fact some of the strong lines of laboratory sources are not observed at all in the nebulae.

Wright sent his measurements to Bowen, who immediately recognized that the lines resulted from the selective excitation of a single high-energy level of O^{++}, followed by subsequent downward transitions from it and the levels it in turn populated. Further, Bowen explained why this single O^{++} level was so strongly populated in nebulae; it resulted from "resonance fluorescence," the absorption by oxygen ions of a strong helium ultraviolet emission line that happens to have exactly the same wavelength as an O^{++} absorption line. Thus Wright's spectroscopic measurements of the nebulae, interpreted by Bowen's physical reasoning, led to new insights into the nature of these objects.

After he became director of Lick Observatory, Wright succeeded in getting funds from a wealthy supporter of astronomy to bring outstanding senior scientists to Mount Hamilton temporarily for work on their research. His plan was based on a similar Research Associate program that had long been in operation at Mount Wilson Observatory, where Henry Norris Russell and Sir James Jeans had gone on frequent research visits. At Lick, Wright invited Ejnar Hertzsprung, the grand old man of statistical astronomy, as the first Alexander Morrison Research Associate in the summer of 1937. The following year Bowen was invited, and he spent the summer of 1938 at Lick Observatory. With Arthur B. Wyse, a young staff member and Wright's protege, he obtained spectra of gaseous nebulae. It was Bowen's first observational work in astronomy, but he already had laboratory skills and with Wyse's coaching on the astronomical details soon became a real expert. New, faster photographic plates were just coming out from the Eastman Kodak Laboratories, and with them Bowen and Wyse were able to detect and identify many new forbidden nebular lines. By carefully calibrating their plates, they were able to get accurate information on line intensities, and from them calculate fairly good relative abundances of the elements in the nebulae. They found the elements to be present in essentially the same ratios as in the sun, with hydrogen the most abundant element, helium second, and oxygen, nitrogen, and neon next. It was an important result, especially as an indication that the stars and nebulae are made of essentially the same stuff.

In the fall of 1939, Bowen returned to Caltech. He had always been interested in astronomy, and this interlude at Lick freshened his interest. It also developed his credentials as at least a part-time astronomer. After World War II Bowen became the first director of the combined Mount Wilson and Palomar Observatories, a post he held for eighteen years until his retirement in 1964.

Wyse, who worked with Bowen at Lick, had originally come to Mount Hamilton as a graduate student in 1931. The son of a Presbyterian minister, he had done his undergraduate work at Wooster College in Ohio, and then had one year as a graduate student of astronomy at the University of Michigan, where he earned his master's degree. Heber D. Curtis, his teacher at Ann Arbor, recognized him as an extremely talented research prospect, and encouraged him to go on to Lick. Wyse was eager to start observational work, and spent the summers and Christmas vacations between his courses at Berkeley on Mount Hamilton. He completed his thesis on the spectra of eclipsing binary stars in 1934, received his Ph.D., married his college sweetheart, and was appointed to the Martin Kellogg postdoctoral fellowship all within a month.

That same year Nicholas U. Mayall also was awarded his doctor's degree. Mayall, a native of Stockton, California, had done his undergraduate work at Berkeley. He started in engineering but soon switched to astronomy, in which he had been interested since his high school science club's visit to Lick Observatory. After graduation he stayed on at Berkeley as an astronomy graduate student for one year, but then in 1929, through working as a grader for Mount Wilson astronomer Seth B. Nicholson, who taught summer school at the University of California that year, Mayall got a job as a research assistant in Pasadena. There he measured and reduced spectroscopic and direct plates for Nicholson, Walter S. Adams, Edwin Hubble, and Milton L. Humason, some of the outstanding research astronomers of their era. In what time he could find after carrying out his assigned measurements, Mayall went to Mount Wilson with these astronomers and learned their observational techniques at the 60- and 100-inch telescopes. He especially enjoyed the work on faint "extragalactic nebulae," which Curtis had said and Hubble had proved were really galaxies, each containing many billions of stars.

After two years at Mount Wilson, Mayall returned to Berkeley for a year's study in formal courses and then went to Lick, where he used the Crossley reflector for his thesis on direct photographs of galaxies. The general subject had been suggested to him by Hubble. Encouraged by Aitken and Wright, Mayall next planned a fast, efficient spectrograph especially designed to be used on the Crossley to measure the radial velocities of faint spiral and irregular galaxies. They hoped to keep him on the staff after he got his Ph.D. degree in 1934, but there were no openings and the Depression had ended any hope of adding new positions. Like Wyse, Mayall got married soon after his graduation, but he had no job and no prospects. At the last moment a janitor at the observatory quit, and Aitken converted his position to an observing assistantship for Mayall for one year. This enabled him to go ahead and start his galaxy observing program with his fast spectrograph on the Crossley.

Mayall and Wyse were both excellent research prospects, intelligent, hard working, and skilled observers. They were also very personable young men. Wright and Aitken desperately wanted to keep them on the Lick staff. But the Depression had dried up all sources of funds, and no new positions could be created. However, Aitken's retirement in the summer of 1935, together with Trumpler's move to the Berkeley campus, despite his official membership on the Lick staff, made it possible to appoint two new assistant astronomers at the junior level. It was just at this time that double-star observer Gerard P. Kuiper's postdoctoral fellowship was ending. He was another excellent prospect, whom Aitken especially wanted on the staff to continue his own double-star observing program. But from the Lick perspective as defined by Wright, there was really no contest. Mayall and Wyse were Lick graduates, Americans, and married, while Kuiper was a Leiden Ph.D., Dutch, and single. It was hard, but he could get a job somewhere else (as he did at Harvard). Mayall and Wyse were appointed to the Lick staff, and went on to do excellent research for the rest of their careers.

Frank E. Ross, another astronomer who was often at Lick Observatory in the 1920s and 1930s, fitted Wright's picture of a "real American." One of the first Lick fellows, he had switched to mathematics for his Ph.D. in 1901, and then after various jobs in astronomy had ended up on the Yerkes Observatory staff in 1924. For nine years prior to that, Ross had worked for the Eastman Kodak Company, and had made himself the outstanding expert in America on astronomical optics. He had designed the best wide-angle lens for astronomical photography then in existence, the four-element "Ross lens." It was far superior optically to the portrait cameras that E. E. Barnard had used for his earlier wide-field photographs of the Milky Way, and with it Ross had produced his new atlas that showed many more stars, nebulae, and dark interstellar clouds.

Ross regarded Yerkes Observatory as an unsatisfactory site for an observatory, and found its cold and moist climate bad for his health and morale. He used it only as his base of operations, and spent as much time as he could observing at the warmer, darker, and clearer Mount Wilson, Lowell, and Lick Observatories. Ross wanted a job at Lick, as he told Aitken and Wright on many occasions, but although they were always friendly and polite in their replies, they naturally preferred young, eager, observational astrophysicists to an old, disaffected classical astronomer whenever they actually had jobs to fill.

Wright at Lick with the Crossley reflector, and Ross at Yerkes and at Mount Wilson, independently pioneered in monochromatic photography of the planets. In 1924 and 1926 Wright obtained particularly good results on Mars, showing that its surface features appear progressively clearer at longer wavelengths and practically disappeared in ultraviolet light. He interpreted these results in terms of scattering in the planet's atmosphere. Ross, using special filters and sensitized photographic plates, as Wright had, obtained some very high-quality monochromatic pictures of Mars in 1926 and of Venus the following year. Always jealous of his own turf, Wright felt that Ross did not give him enough credit for his earlier work, and they clashed gently by correspondence, but soon became reconciled.

Through his lens design, Ross was involved in the great project that became Wright's legacy to Lick Observatory. This was the Lick proper-motion survey, based on an idea that Wright had originally had in 1916. He and other Lick astronomers were then beginning to realize through Curtis' observational research, that spiral "nebulae" were in reality extremely distant galaxies. Compared with stars, faint galaxies were essentially at infinite distance, and thus were ideal markers (a reference system, in astronomical terminology) against which to measure the small angular motions (proper motions) of the stars. Furthermore, the universe was actually composed of the galaxies, so the reference system they define is, in terms of physics, the natural one with respect to which the stars' motions *should* be measured. All previous measurements of stars' proper motions had been made with respect to other stars, and ultimately, through comparison with measurements of planets or the sun, with respect to a reference system fixed in the solar system. Because of the slow precession of the earth's axis in space and other smaller effects, it was very difficult to convert the measured proper motions in this reference system to the desired proper motions in the ideal reference system fixed in the universe.

These traditional measurements made visually with a meridian circle, a small, specially mounted telescope, had been a minor part of the Lick observing program from the beginning. John M. Schaeberle of the original Lick staff had begun them, and his successors Richard H. Tucker and Hamilton M. Jeffers had continued them, but none of their work had ever had much impact on the wider field of astronomy.

Wright realized that he could bypass the meridian-circle measurements with respect to planets, and measure the proper motions of many more stars directly with respect to galaxies in the ideal reference system. However, to do so required long-exposure wide-field photographs, each of which would show the images of hundreds of galaxies. Wright's experiments with the Crossley reflector during World War I convinced him that a conventional reflecting telescope could not be used for this program; it had too small a usable field and the number of exposures required to cover the sky would have been prohibitive.

After Ross had developed and demonstrated his new lens system in 1928, Trumpler and Wright both became interested in acquiring one for Lick Observatory. Trumpler wanted it for his studies of star clusters, and Wright for the proper-motion program. However, the special glasses required to make the lens were expensive, money was hard to come by, and the idea languished for several years. Trumpler gradually lost interest in it, but Wright did not. In 1934, during the depths of the Depression, he and Director Aitken finally succeeded in getting a grant of $65,000 from the Carnegie Corporation for a 20-inch astrographic telescope. Ross provided the optical design and the lens was ordered from J. W. Fecker, the successor to John A. Brashear and by now the main American supplier of astronomical optics. During Wright's directorship, building this astrograph was his main research project. He outlined the principal specifications for the mounting, which the Warner and Swasey Company then designed in detail. Wright travelled often to Fecker's shop in Pittsburgh and the Warner and Swasey plant in Cleveland to consult and check on progress.

Wright was the unofficial mayor of Mount Hamilton, just as Holden, Keeler, Campbell, and Aitken had been before him. In the 1930s, the decade of the Great Depression, the mountain's population slowly dropped from fifty to forty. This partly mirrored the falling birthrate throughout the country, but the departure of the Trumpler family for Berkeley also played a significant role in the decrease. In 1936 the Mount Hamilton school graduated its last two eighth graders, and had to close its doors. Still, the "stage," a small truck or van that carried passengers as well as freight, came up Mount Hamilton every day but Sunday. Its regular driver, Ernest Roper, had been on the job nearly twenty-five years in 1942, and was known to a generation of graduate students and assistants, whom he had ferried up the mountain, along with milk, butter, eggs, and other food supplies ordered by the resident astronomers' wives. The two stops along the Mount Hamilton Road where the old stage had changed horses were no more; the Grand View Hotel, overlooking the Santa Clara Valley, burned down in 1942, and the Smith Creek Resort was razed in 1937. According to an observatory legend there were exactly 365 turns in the original road in the seven-mile climb from Smith Creek to the summit. After the road was improved in the 1930s, only 287 turns were left according to Roper, but he still had to negotiate them all, including "Devil's Elbow," the "Horseshoe," and "Oh! My," named in the early days for the passengers' exclamations the first time they saw it.

During the years just before America entered World War II, three more bright young research workers completed their graduate training at Lick Observatory. They were Daniel M. Popper, Horace W. Babcock, and Gerald E. Kron, who all received their Ph.D. degrees in 1938. For his thesis Babcock measured the rotation of M 31, the Andromeda galaxy, using the fast nebular spectrograph Mayall had developed on the Crossley reflector. Babcock obtained several exposures with this spectrograph in conjunction with a 5-inch parabolic mirror, used as a very small telescope. It allowed him to focus the image of a large part of the galaxy onto the slit of the spectrograph, and record its velocity curve on one exposure.

Kron came from the University of Wisconsin, where he had started in engineering and then had earned a master's degree in astronomy working with Joel Stebbins. Kron brought Stebbins' photoelectric techniques with him, and his Lick thesis on photometric measurements of eclipsing binary stars established new standards of precision. From them he drew valuable physical information on these star systems.

World War II was brought home to Lick Observatory even before the United States got into it. California was a major center for aviation training, and on May 21, 1939, an Army Air Force A-17 attack bomber, flying blind in the fog, crashed into the Main Building on Mount Hamilton. If it had been fifty feet higher, it would have gone safely over the building; if it had been a hundred and fifty feet to the right, it would have struck the dome of the 36-inch refractor. Instead, it crashed straight into the Main Building, just at the first-floor level. It tore through two brick walls and came to rest in the corridor, a tangled mass of crushed fuselage and wing, bricks and mortar. Both crew members were instantly killed. It was seven in the evening, just supper time, and only one person was in the building; otherwise anyone who had been in the offices through which it crashed probably would have been badly injured or killed. The fuel from the plane poured into the building, but luckily did not ignite. None of the astronomical instruments was damaged. After the wreckage had been removed, the walls were rebuilt, but the change in the brickwork can still be seen.

As World War II approached, the younger Lick astronomers one by one left the observatory to apply their scientific training and experience to the war effort. Wyse, who had been working with Wright on testing and aligning the 20-inch Ross lens of the astrograph, joined the Navy Radio and Sound Laboratory in San Diego the week before Pearl Harbor. It was the main center for the development of antisubmarine detection systems. Wyse threw his heart into his new assignment, but in his evening spare time managed to complete a paper on his spectroscopic measurements of gaseous nebulae. Less than six months after leaving Mount Hamilton, he went east to the navy's lighter-than-air base at Lakehurst, New Jersey, for field tests of a dirigible-borne sub-marine-location system. There, on the night of June 9, 1942, Wyse was killed together with eleven other civilian scientists and naval aviators in the collision of two blimps over the Atlantic Ocean. Wright, terribly shaken, wrote that "[t]he Lick Observatory has suffered a blow from which it will not recover in our time." Wyse was dedicated to research, highly intelligent, and mechanically very adept. He was an excellent teacher, and extremely personable. Everyone who knew him believed that he would some day be director of Lick Observatory, and no doubt he would have been, if he had survived World War II.

The first Lick staff member to go on leave for war service had been the recently appointed Kron, who went to the Radiation Laboratory on the Massachusetts Institute of Technology campus in May 1941. It was the center of radar development in the United States, and Kron's skills in electronics made him well qualified for the work there. In October Jeffers joined him at the Radiation Laboratory. Before long the older astronomer, who was an amateur pilot, became an expert in operations analysis, the application of scientific methods and statistical concepts to the tactics of air warfare. Jeffers served for six months in 1943 as a civilian technical adviser with the 11th Air Force, stationed at Adak in the Aleutian Islands. Then he became the head of the Operations Analysis Section for the B-29 squadrons that trained in Kansas, and initiated long-range bombing of Japan from bases in the interior of China in 1944.

In June 1942, six months after America's entrance into Warld War II, Mayall became the third Lick astronomer to join the Radiation Laboratory staff. However, the raw New England winter proved too severe for him, and he suffered constantly from colds, arthritis, and other ill-nesses. In the summer of 1943, he transferred to an army optical project at the Mount Wilson Observatory headquarters in Pasadena, designing and testing aerial cameras, gunsights, range finders, and the like. Early in 1944 he moved on to join a photographic group, headed by Bowen, in the rocket development program at the California Institute of Technology.

World War II brought other changes in the operations on Mount Hamilton. One was the suspension of the popular Saturday evening visitors' program, which had been in existence since the observatory's completion in 1888. Immediately after Pearl Harbor, military authorities feared that Japanese submarines, lurking off the coast of California, might use the headlights of visitors' cars, streaming to the summit of Mount Hamilton, as a navigational marker. After the war ended the popular visitors' nights were reestablished on Friday evenings in the summer, rather than Saturday evenings throughout the year.

After the war had begun, Wright continued as director until June 30, 1942, when he stepped down at the age of seventy. However, because of the shortage of younger astronomers, he continued as a staff member under a special appointment for two more years. When he finally left the mountain in 1944, he had the Carnegie astrograph assembled, tested, and essentially ready for use.

Joseph H. Moore succeeded Wright as director. Sixty-three years old when he stepped into the post, Moore was appointed as interim director for the duration of the war. He was the last of the old guard who had been at Mount Hamilton since nearly the beginning of the observatory. Born in Ohio, he attended Wilmington College and became interested in astronomy there. After receiving his A.B. degree in 1897, he had gone on to Johns Hopkins University to study as a graduate student under Simon Newcomb. Moore's preparation was woefully inadequate, and he had to spend his first two years taking undergraduate mathematics and physics courses. During that period Johns Hopkins dropped its graduate department of astronomy, and Moore changed his major to physics.

He worked in laboratory spectroscopy under Henry Rowland, and after the latter's death under his brilliant successor Robert W. Wood. Johns Hopkins was the premier center for training in physics in the United States, and when Moore completed his Ph.D. in 1903, Joseph S. Ames, the chairman of the department, called him in and told him that for his first job he could choose between requests for instructors at Harvard, Yale, and the University of California, and for an assistant in spectroscopy at Lick Observatory. The starting salary at Lick was the smallest of the four, but Moore, still interested in astronomy, chose it. He spent the rest of his life on the Lick staff.

When Moore arrived at Mount Hamilton, he was the only staff member with an earned Ph.D. degree. (Henry Crew had been the one previous Ph.D. hired at Lick, but lasted only a year on the staff.) After Moore, practically no one was hired without a doctor's degree. Wright was about to leave for Chile to start the Lick southern observatory, and Moore replaced him as Campbell's assistant on the radial-velocity program. In 1907 he married Fredrica Chase, a Vassar graduate who had come to Mount Hamilton as a computing assistant in 1905. She retained her interest in astronomy, and when Moore went to Santiago to head the Lick southern station from 1909 until 1913, she helped him with measuring and reduction work there. After his return to Mount Hamilton, Moore took over more and more of the responsibility of running the radial-velocity program, and when Campbell became president of the University of California in 1923, Moore was left in effective charge of this research. He participated in five eclipse expeditions, the first two with Campbell, at Goldendale in 1918 and Wallal in 1922, and he did some spectroscopic work on nebulae and planets. But he spent most of his career in measuring the radial velocities of stars.

During Moore's wartime directorship most of the regular staff members were gone to war work, and research was at a low ebb on Mount Hamilton. However, two very important figures in the later history of Lick Observatory began their observational work there. One was Lawrence H. Aller, the other, George H. Herbig.

Aller, born in Tacoma, Washington, was interested in astronomy almost from birth. However, in 1929, when he was 15 years old, his family was desperately poor, and he had to quit high school after only two years and go to work with his father in a gold mine that never proved successful. Aller kept studying and reading on his own. He joined the Astronomical Society of the Pacific later that year, "squandering" (in his father's words) a gift of $3.00 one of his older brothers had sent him on a year's membership dues. Aller struck up a correspondence with Donald H. Menzel, then on the Lick staff, and in November 1931 managed to meet him on the Berkeley campus, where the astronomer was teaching that semester. Menzel let young Lawrence take the final examination he was giving to his elementary astronomy class. The eighteen-year-old high school dropout did so well on it that Menzel managed to convince the University of California administration to admit him as a special student. Aller quickly made up his educational deficiencies and earned his A.B. in 1936. He stayed on as a graduate student in Berkeley for one year, and in the summer of 1937 he worked as an assistant for Wyse and Mayall on Mount Hamilton, and made observing charts for Hertzsprung.

By then Menzel had moved to Harvard, and Aller, who had a brilliant academic record at Berkeley, followed him and was admitted as a graduate student there in the fall of 1937. He wanted to study astrophysics, rather than the traditional astronomy that still dominated Lick. However, observing opportunities were limited at Harvard but not at Mount Hamilton, so Aller was glad when Wright hired him again as a summer assistant in 1938. The next year, on the recommendation of Menzel and Wright, Aller was appointed a junior member of the Harvard Society of Fellows, a very prestigious group. With this fellowship he was able to come back for more observational work on gaseous nebulae with the Lick Crossley reflector in the autumn of 1939, and again for a few more nights during his summer vacation in 1940.

Aller received his Ph.D. at Harvard in 1943, and then came west to the Radiation Laboratory at the University of California campus in Berkeley, where under director Ernest O. Lawrence he was a member of the team that developed the electromagnetic method ("calutron") for

230   THE LAST OF THE FIRST

separating uranium isotopes for the atomic bomb. But whenever Aller could get away on a weekend, he would somehow manage to find a ride up to Mount Hamilton for a few nights of spectroscopic observing of gaseous nebulae with the Crossley reflector. Much of Aller's observational material for his early studies of the abundances of the elements in planetary nebulae was obtained in those wartime years.

Herbig, seven years younger than Aller, was also a dedicated astronomer from his earliest years. He grew up in Los Angeles, where he was a part-time guide at Griffith Planetarium, and visited Mount Wilson Observatory where he was inspired by astronomer Alfred H. Joy. Herbig majored in astronomy at the University of California at Los Angeles, where he finished his undergraduate work in 1943. Like the older Aller, he went immediately to war work at the Berkeley Radiation Laboratory, but his health was poor and he was not strong enough for the job. He was released by the Radiation Laboratory and went to Lick as one of the few graduate students and assistants of that period. Both he and Aller were dedicated observers and intelligent, creative scientific thinkers. Both were to go far in astronomy.

Finally in the summer of 1945, World War II ended with a sweeping Allied victory. Heroism in the field combined with scientific developments from the laboratories had defeated first Germany and then Japan. The Lick astronomers started coming home to a new era of astronomy that was about to begin on Mount Hamilton.

13

Second Largest Telescope in the World
1945–1958

On July 16, 1945, a brilliant burst of light flashed out above the New Mexico desert. Windows rattled in Albuquerque, in El Paso, and even in Gallup, over two hundred miles from the Trinity site where American scientists test fired the first atomic bomb in the history of mankind. It was an awesome sight; its destructive powers surpassed even the most optimistic estimates of its designers.

Three weeks later a car headed down the long grade from Los Alamos to Santa Fe, continued south to Albuquerque, and turned west toward California. At the wheel was C. Donald Shane, assistant director for scientific personnel of the laboratory that had built the test bomb, the "Little Boy" uranium bomb that was dropped on Hiroshima the day before he left, and the "Fat Man" plutonium bomb that exploded over Nagasaki while he was on the road. Japan had sued for surrender before he reached his home in Berkeley. Shane was privy to all the secrets of the atomic bomb project, and after he had witnessed the Trinity test he had been confident that the war would soon be over. He was going back to the University of California to become director of Lick Observatory.

Shane was born on a ranch near Auburn, California, where his father was the principal of a rural grammar school. He attended the University of California, where he earned his undergraduate degree in 1915, spent one year as a graduate teaching fellow in mathematics, and then became a graduate student in astronomy. After two years at Lick Observatory, separated by two years teaching navigation during World War I, he earned his Ph.D. in 1920. His thesis, done under the supervision of W. W. Campbell, was an observational study of the spectra of carbon stars, relatively rare objects in which helium and carbon are unusually abundant.

Immediately after receiving his Ph.D., Shane became an instructor in mathematics at Berkeley. He gradually transferred his activities into astronomy, and was promoted to assistant professor of astronomy in 1924. He worked his way up to professor in 1935, and chairman of the astronomy department in 1941. Although Shane did some research, his main contribution up to the time of World War II was thus in teaching. He taught astrophysics to nearly all the advanced astronomy undergraduates and graduate students in Berkeley between 1924 and 1942.

In the summer after Pearl Harbor, Shane's friend, physicist Ernest O. Lawrence, asked him to join the staff of his Radiation Laboratory in Berkeley, where scientists were developing the electromagnetic process for separating the isotopes of uranium. Shane accepted and became Lawrence's assistant director for scientific personnel. In 1944 he moved on to Los Alamos in the same position under J. Robert Oppenheimer, another Berkeley physics professor whom he knew well.

Shane was a friendly, diplomatic, hard-working professor. He took a very active part in the Berkeley faculty senate, and at one time or other served on nearly every important faculty committee there. He had become well acquainted with all the important figures on the Berkeley campus, and was a close personal adviser of Robert G. Sproul, who was appointed as president of the University of California after Campbell retired in 1930. Shane was to succeed, where all the previous Lick directors had failed, in seeing a large, modern reflecting telescope built at Mount Hamilton, second only to those of Mount Wilson and Palomar Observatories to the south.

The founder of those observatories, a figure whose career was entwined with the history of Lick Observatory for over forty years, was George Ellery Hale. Born in Chicago, the scion of a very wealthy family, Hale had first visited Lick Observatory on his honeymoon in 1890, immediately after his graduation from the Massachusetts Institute of Technology. At Mount Hamilton Hale discussed eclipse photographs of the corona with John M. Schaeberle, and at night watched James E. Keeler observe with the 36-inch telescope, then the largest telescope in the world. Even years afterward Hale could not forget that first night in the dark of the great dome, with the long tube of the refractor reaching up toward the slit, which seemed to him an opening into heaven.

A little over two years later Hale, by then a faculty member at the University of Chicago, and its President William Rainey Harper succeeded in persuading Charles T. Yerkes to provide the money for a 40-inch refractor. The Chicago newspapers boasted that it would "lick the

Lick." Hale brought S. W. Burnham back from retirement to part-time observational work on the Yerkes staff, and lured E. E. Barnard away from Lick to join him. He very nearly added his close friend, James E. Keeler, to the Yerkes faculty also, but Keeler accepted the Lick directorship instead.

Hale, always more interested in building telescopes than in using them, had tired of the 40-inch before it was completed. He wanted to erect a larger reflecting telescope in the clearer, more stable skies of California. Hale's wealthy father bought a 60-inch glass disk for him, and provided the money to hire George W. Ritchey to begin grinding it into a mirror even before the 40-inch refractor had gone into operation. At first Hale hoped to erect the 60-inch reflector in California as a remote station of Yerkes Observatory, but Harper had other priorities and would not allot the necessary funds. Andrew Carnegie, the steel magnate, was just beginning to think about giving away his vast fortune, and in 1902 Hale succeeded in getting himself appointed to a committee to advise him on how he could best help astronomy. Within a few months Carnegie granted funds for Hale, Campbell, and Lewis Boss of Dudley Observatory in Albany, New York, to investigate the possibility of setting up "a southern and solar observatory." Campbell, as the resident California expert, sent William J. Hussey on a site survey in southern California. He recommended Mount Wilson, in the Sierra Madre range above Pasadena, a site which Hale had known and coveted for years. He and Campbell first visited it, with Hussey, in June 1903. Campbell hoped to get a "four-foot" (48-inch) reflector for Mount Hamilton with Carnegie's money, but Hale was much more skillful in presenting his own case to the wealthy man and his advisers. The result was the establishment of the Mount Wilson Solar Observatory, with Hale, who soon severed his ties with Yerkes Observatory, as its director, and money to erect the 60-inch telescope, but no funds for a 48-inch for Lick.

By the end of 1908 the 60-inch reflector was in operation on top of Mount Wilson, and Lick Observatory's telescopic equipment had been surpassed, as it never had been by the Yerkes 40-inch in the cloudier midwestern skies. Hale was the all-time great promoter and fund raiser of astronomy, and within a decade he had the even more powerful 100-inch reflector in operation beside the 60-inch atop Mount Wilson. It was the unchallenged world center of observational astronomy. Campbell had originally introduced Hale to John D. Hooker, the Los Angeles hardware magnate and amateur astronomer who supplied the initial grant of money that got the 100-inch project started.

Hale, a superb diplomat, always kept on good terms with the Lick astronomers. For thirty years he and Campbell were dominant figures in U. S. astronomy. Highly neurotic and frequently subject to bouts of mental illness in which he would withdraw to sanitariums or on long travels to take him away from scientific research, Hale encouraged the older, more stable Campbell to take the lead in their joint projects. Hale, more adept in manipulating people, suggested many of the organizational ideas, committee appointments, and prize awards in American astronomy, which Campbell helped him put into effect.

Several of the early Mount Wilson astronomers received their training at Lick before they moved on to the big telescopes in the south. Among them were Paul W. Merrill, Seth B. Nicholson, and Roscoe F. Sanford. At Hale's invitation William H. Wright, who had worked with him for a year at Kenwood and Yerkes Observatories, brought his quartz spectrograph to Mount Wilson and used it as a guest observer with the 100-inch telescope, soon after it went into regularly scheduled operation. Thus although the Lick telescopes were definitely second best, Hale never emphasized Mount Wilson's superiority, and always encouraged the Mount Hamilton observers to pride themselves on what they were able to accomplish with their increasingly antiquated equipment.

Every Lick director from Campbell onward tried to acquire a more powerful telescope for the observatory. In 1902, Campbell had proposed to the Carnegie advisers first a 54-inch reflector, then a 48-inch reflector, to be erected in Chile at the Lick southern station, but Hale's Mount Wilson plan had won out. Probably if Campbell had not become president of the University of California in 1923 he would have succeeded in securing a large telescope for Lick Observatory. When he returned from the Wallal eclipse in 1922 he was at the peak of his fame and prestige. America was prosperous, and if Campbell had devoted his efforts to cultivating California millionaires, he probably would have succeeded in raising the money necessary to erect a large reflector on Mount Hamilton. Instead, he accepted the presidency and all his time and energy went into his new job.

Robert G. Aitken, as associate director, considered it his duty to keep the observatory running smoothly, but not to seek outside financial support. Only in 1930, after Campbell retired and Aitken became director, did the two of them approach the Carnegie Corporation and the Rockefeller Foundation to fund a 72-inch reflector for Lick Observatory. It was too late. Carnegie was already committed to Hale and Mount Wilson Observatory, and in 1928 the aging supreme promoter of astronomy had pulled himself together, emerged from seclusion, and persuaded Rockefeller's advisers to commit their patron's money to the 200-inch Palomar telescope. By 1930 the Great Depression had arrived and there was no more money to be had; in 1932 when Aitken tried to get the funds for the proposed telescope from local California sources the situation was even more hopeless.

When Wright became the director of Lick Observatory, he also tried to raise funds for a 72-inch reflector, first from the Rockefeller Foundation and wealthy Californians, then from within the University of California. He was equally unsuccessful in both attempts. America was still in a depression in 1941, the last time Wright appealed to President Sproul, and there was no tradition of spending state tax funds on expensive capital costs of research equipment. By then the estimated cost of the telescope, dome, and necessary auxiliary instrumentation had grown to $450 thousand.

Aitken and Wright had both reported in their official communications to Sproul that the Lick telescopes were antiquated, and that the observatory could not go on as a first-class research institution indefinitely without a larger telescope. No doubt Shane, in his informal conversations with the president, emphasized exactly the same point. But none of the Lick directors ever thought of trying to top Hale; from Campbell's 54-inch to Wright's 72- to 75-inch, they always proposed telescopes that would leave Lick in the second position, behind Mount Wilson. In spite of the astronomers' recommendations, Sproul and the Berkeley physicists who advised him urged the successive Lick directors to emphasize astrophysics and emulate Mount Wilson, but did not provide the money for a telescope that would enable them to do so.

However, in late 1941 Sproul finally saw the light. A rejection at last confirmed the harsh reality of the arguments to which he had been subjected over the years. In 1939 he had set up a committee, chaired by physicist Raymond T. Birge, to advise him on whom to appoint to succeed Wright, scheduled to retire as director of Lick Observatory in June 1942. The other members of the committee were Aitken, by then director emeritus and living in Berkeley, R. Tracy Crawford, Robert J. Trumpler, and Shane of the Berkeley Astronomy Department, plus another physicist, a chemist, and a mathematician, all from the Berkeley faculty. Their first choice for the directorship was Merrill, the Lick Ph.D. who had joined the Mount Wilson staff and become an outstanding stellar spectroscopist. However, when he was sounded out informally, Merrill firmly stated that he was not interested in leaving the 60- and 100-inch telescopes with the 200-inch also nearing completion, even to become director of Lick Observatory, if he would only be able to observe with the 36-inch refractor and the Crossley (36-inch) reflector. From that time onward, Sproul was convinced that Lick must have a new large reflector after World War II ended.

The president met with the Lick Observatory Committee of the regents three months after Pearl Harbor, and they unanimously agreed they could not attempt to raise the funds for a telescope to lure Merrill to Mount Hamilton with the war in progress. Instead they decided to offer the directorship to "the best man available," . . . "a younger man of promise" who would be willing to take over the observatory as it was, and emphasize the types of research that could be continued with its smaller telescopes. He would also be expected to continue and strengthen the graduate astronomy training, which they all considered the best in the country.

The man they had in mind was Shane, who had been the second choice of nearly all of Birge's committee members, but by the time Sproul could offer him the position, it was clear that neither he nor anyone else could accomplish much at Lick until after the war had ended. Shane was already involved in teaching special wartime courses on the campus, and he knew that he would soon be drawn into even more important projects. He advised Sproul that he could not take the Lick position, and recommended that instead he appoint Joseph H. Moore, then sixty-three years old, as interim wartime director. He had been the one candidate whom all the committee members had considered too old for the job, but by 1942 he was the only choice left. A month later Lawrence offered Shane the assistant directorship at the Radiation Laboratory, working on the country's highest-priority scientific development.

Sproul appointed Moore the director of Lick Observatory, but he also put into the university's ten-year statewide building program an item of $600 thousand for a reflecting telescope for Mount Hamilton. It was the first project scheduled for the second of the ten years, which clearly could only begin after the war. The position of this item in the ten-year $54 million program for all the campuses indicated the great importance Sproul attached to it. The cost figure for the telescope, of unspecified aperture, was a scaled-up version of Wright's estimate, and the paragraphs explaining the need for it were taken from his report. All the items in the program were still tentative, however, until approved and funded by the legislature and the governor.

Moore, soon after taking over the directorship, wrote to Sproul to emphasize again how important the telescope would be for the future of Lick Observatory. By then the interim director visualized it as an 80- or 85-inch reflector. In the dark days of World War II the eventual construction of the telescope seemed far in the future, but by the fall of 1944, the Allied armies were lodged on the continent of Europe, had liberated Paris, and were pressing toward the borders of Germany. Victory seemed in the air, and the astronomers' thoughts returned to a postwar telescope. By this time Sproul had raised the item in the postwar building program for the Mount Hamilton reflector to $900 thousand, and Moore had revised his thoughts upward to making it 90- or 100-inch.

However, Lick staff member Nicholas U. Mayall, on leave for war work at the California Institute of Technology, had other ideas. In his student days, he had been an assistant at Mount Wilson for two years, and was a strong convert to its big-telescope philosophy. At the Caltech optical shop he saw the large 120-inch diameter glass disk that had been intended for use as a "flat," or plane reflecting mirror, for testing the 200-inch mirror as it was ground, figured, and polished to paraboloidal shape. Why not use that surplus disk, and make the Lick telescope a 120-inch reflector, Mayall thought. He discussed his idea with fellow Lick astronomer Gerald E. Kron, also on leave for wartime technical research at Caltech. Kron was enthusiastic.

Mayall then arranged for the two of them to see President Sproul in mid-December 1944, on one of his regular visits to the UCLA campus. Mayall did most of the talking. He emphasized that he, Kron, and the other younger members of the staff would be the actual users of the telescope, not the older, more conservative Moore and Wright. He urged that the telescope be built as large as possible, and emphasized the availability of the 120-inch disk. The University of California should not "sell out" for a moderate-sized instrument. Sproul agreed and encouraged Mayall and Kron to continue to think big.

Three months previously Shane had been back in Berkeley for a brief visit and had met with Sproul. No doubt they had both understood since 1942 that as soon as the war was over Sproul would appoint Shane to succeed Moore as director of Lick. Now, with the war apparently nearing its end, the president got the regents' approval to make the appointment at once, secretly, with the condition that it would become effective just as soon as Shane could return. They also approved Sproul's commitment to Shane to provide a modern reflecting telescope with a diameter of 85-inches or more, plus auxiliary instrumentation, at a total cost of $900 thousand.

In the meantime the president decided to set up a committee to advise him on the necessary first decisions as to the telescope size, form, and location. From the beginning Shane wanted to use as much of the Mount Wilson and Palomar expertise and experience as possible, preserving the University of California money for construction of the telescope. The committee, as appointed by Sproul following his recommendation, consisted of Shane as chairman, Moore, Mayall, Walter S. Adams, director of Mount Wilson Observatory, and Ira S. Bowen, the Caltech physicist who had done so much for astronomy and who was to succeed Adams in 1946.

Shane's idea was to build as large a telescope as they could for the $900 thousand; he thought it would be about a 90-inch. He wanted it to be a general-purpose telescope, with the capability of doing high-dispersion spectroscopy, the astrophysical work that was Mount Wilson's specialty and that Sproul, Birge, Lawrence, and all the Berkeley physicists appreciated so much. At first Shane thought this would require that the telescope have an expensive auxiliary coudé focus, at the lower end of the polar axis, with a fixed coudé spectrograph in a room far below the observing floor. Moore favored omitting the coudé to save money, and doing all the spectroscopic work at the Cassegrain focus, an arrangement fairly similar to the 36-inch refractor with which he was familiar. He thought they would be able to afford a 90-inch or 100-inch. Mayall preferred a bare-bones telescope, as large as possible, and well equipped at the prime focus for the nebular and galaxy research in which he specialized. He was willing to sacrifice almost everything else, but insisted that the telescope should be a 120-inch. Sproul authorized Shane to go ahead on the basis of a telescope that would cost $900 thousand at 1940 prices, or in other words, not to worry about postwar inflation. To the cautious Moore this meant they might go to 110-inches, but no higher; to the ebullient Mayall it meant they would have their 120-inch.

Shane was able to get away from Los Alamos for a meeting of the advisory committee in Pasadena on March 6, 1945. Moore had an infected foot and could not travel but Adams, Bowen, and Mayall were there, along with John A. Anderson, the head of the Palomar project, who attended the meeting as a consultant. On that day they made all the basic decisions on what would years later become the Shane 120-inch reflecting telescope. They recommended it have a focal ratio F/5, considered ideal for photography by the Mount Wilson astronomers. It would have a fork mounting of the type originally proposed by Francis G. Pease of Mount Wilson in his early design study for what ultimately became the 200-inch. The Lick telescope would be built for use at the prime focus and Cassegrain, but with a provision for adding the coudé later if desired. And they stated that they believed that a 120-inch telescope could be built with the funds Sproul had promised, the equivalent of $900 thousand in 1940 costs.

That evening Shane and Mayall went on to San Jose by train, and on the next day met Moore on Mount Hamilton. He agreed with their decisions. Wright joined them, and together they picked out a site for the telescope. No one had ever seriously thought of any location but Mount Hamilton. The specific place they picked was on the flat area where the tennis court and Aitken's old house then stood, about a half mile east of the Main Building. Wright proved quite obstructive. By this time his main interest was the 20-inch astrographic telescope he had succeeded in having built for the Lick proper-motion survey. He believed that all the observatory's resources should be devoted to it, and feared that the proposed large telescope would be a diversion from the proper-motion program. Deeply conservative, Wright still believed that a 72-inch would be large enough, and criticized Shane as "naive and childlike" for thinking that a 120-inch could be a success. However, he was the voice of the past. Mayall dreamt of working at the prime focus of a 120-inch telescope, and thought that they could "squeeze the last photon of light from the mirror" and "give the 200-inch a good race with a 120-inch."

By this time, however, Mayall had learned of the problems that the ribbed structure at the back of the 200-inch mirror, designed to save weight, was giving in practice. He recommended that the 120-inch be made from a solid, relatively thick glass disk, rather than from the thinner, ribbed test disk that was available in the Caltech shop. Shane added this recommendation in the final report to Sproul, and the president accepted all the committee's advice. Adams, always conservative, privately cautioned Shane that the $900 thousand might not be quite enough for a 120-inch, but he was sure that it would buy at least a 100-inch.

The very day the committee picked the site for the telescope on Mount Hamilton, American troops crossed the Rhine River at Remagen and began to pour into Germany. As one victory followed another, the astronomers' thoughts turned increasingly toward postwar research. Thus just as soon as Shane could see that the war would end and that he was no longer needed at Los Alamos, he resigned and hurried back to California. Moore had heart trouble and was equally anxious to leave Mount Hamilton. By August 15, 1945, the day after Japan accepted the terms of surrender dictated by the Allies, Shane was on the mountain conferring with Moore. The regents reconfirmed Shane's appointment openly in October, and on November 1 he moved up to Mount Hamilton. On December 1 he formally took over. He was the first director, except for Edward S. Holden and James E. Keeler, who was not promoted directly from the Lick staff, and Keeler had been a staff member for five years before going away to Allegheny.

Even before Shane became director, Mayall, Kron, and Hamilton M. Jeffers had all returned to Mount Hamilton from their wartime posts. Just before Christmas the regents formally adopted the statewide building program, in which the Lick Observatory reflecting telescope was listed as the twentieth of the sixty-two "minimum urgent building needs of the University to 1950," whose costs totalled $60 million. By now the appropriation for the telescope had risen to $1.2 million, reflecting

President Sproul's allowance for inflation and for the conservative planning of the Lick astronomers. The bill came before the legislature in March 1946. Shane was called to testify, and took Mayall with him to Sacramento. They found themselves in the middle of a political skirmish between Governor Earl Warren and a conservative group of legislators. In spite of Shane's explanation of Lick Observatory's need, the telescope was deleted from the bill in committee. However, the legislators, who had only wanted to "teach the governor a lesson," then put the telescope appropriation back in the bill on the floor. They passed the appropriation bill and the governor signed it. At last the project could go ahead. The Lick astronomers were jubilant, and Mayall's friend at Mount Wilson, Milton L. Humason, expressed the general feeling when he wrote that the big reflector plus $100,000 that was also appropriated for rehabilitation of the other telescopes and instrumentation would "without question put Lick Observatory on the map again and with a bang."

Even before he had moved to Mount Hamilton, Shane had gone to Pasadena to consult with Adams, Bowen, Anderson, and Bruce Rule, the chief engineer of the 200-inch project, as to how to proceed and whom to hire. They advised him to try to get Mark Serrurier to design the Lick telescope. He was the principal designer of the 200-inch and the inventor of the concept of balancing flexure and thus minimizing its effects, rather than attempting to eliminate it. Serrurier was committed to other projects but he agreed to act as a consultant; instead of him Shane hired as designer and ultimately chief engineer Wilbur W. "Bill" Baustian, who had worked with Rule toward the end of the Palomar project.

By the fall of 1946 Baustian had essentially completed the preliminary design. He was able to keep the fork mounting that was so attractive to the Lick astronomers, with enough room for a large Cassegrain spectrograph behind the primary mirror. The prime focus was to be used for direct photography and photometry, and the coudé for high-dispersion spectroscopy. Shane estimated that if construction could begin in 1947, the telescope could be completed by 1952.

However, the rapid postwar inflation was much worse than Sproul or Shane had expected. As costs of material and labor increased, it became obvious that something would have to give in the design. By now all the Lick astronomers were committed to keeping the aperture of the planned telescope at 120-inches. The quotations for the price of a glass disk that size from the optical companies turned out to be much higher than their early estimates. To cut costs Shane decided to buy the 120-inch thin, ribbed flat from Caltech, instead of a more expensive thick, solid disk. The Cassegrain focus was eliminated from the design for the present, though provision was left for adding it later if more money became available. For a time the coudé was nearly eliminated also. Mayall considered the prime focus, which he would use himself for direct photographs and low-dispersion spectra of faint galaxies, as essential, but claimed that "even such enthusiastic disciples of high dispersion as W[alter] S. Adams and P[ol] Swings have told me there probably are enough coudé foci currently available to keep spectroscopists busy for quite some time."

The contract for the mounting was eventually given to Judson Pacific-Murphy, a local steel construction company that had never made a telescope before, but that turned out to do a good job on this one. Though Moore had recommended having an outside company provide the mirror, this was judged too expensive and it was made in-house instead. An optical shop was set up in the completed 120-inch dome so that the mirror could be tested in the telescope on stars during the final stages of figuring and polishing, as had been done at Palomar. Donald O. Hendrix, who had finished the 200-inch mirror, was engaged as a consultant to supervise the optics of the 120-inch, and Howard Cowan, who had trained with him, was hired to do much of the work at Mount Hamilton.

Part of Shane's policy as director was to encourage the closest possible cooperation with the Mount Wilson and Palomar Observatories staff. Mayall was an enthusiastic follower of this concept; he worshipped Edwin Hubble, the great observational cosmologist, and thought very highly of Humason and Walter Baade. All three of them had done great things with the 100-inch, and were opening new horizons in the study of galaxies with the 200-inch, after it went into regular operation in 1950. Shane arranged for frequent meetings of the "nebular" (galaxy) research workers of the two observatories.

Mayall, an assiduous and extremely skilled observer, used the Crossley telescope to its best advantage while waiting for the 120-inch to be completed. In 1934, he had designed and had constructed an extremely fast spectrograph that was optimized for work on faint galaxies and nebulae. He arranged his program to concentrate on extended irregular and spiral galaxies, for which the aperture of the telescope is less important than it is for the elliptical galaxies, with their concentrated, almost stellar nuclei, which Humason observed at Palomar. Ultimately, their results, along with photometric measurements made with the 200-inch by Allan R. Sandage, were published in a very important joint paper in 1957. For many years it remained the most significant observational discussion of the expansion of the universe.

Kron, who had begun his training in photoelectric photometry with Joel Stebbins at Wisconsin before earning his Ph.D. at Lick in 1938, became one of the world experts in this branch of research in the post-war years at Mount Hamilton. He was one of the first to use photo-multipliers, with their very high sensitivity, for astronomical measurements. Kron's work on eclipsing variable stars was important in establishing their dimensions, temperatures, and other properties, including the fact that some cool dwarf stars had "starspots" analogous to sunspots. Stebbins by this time had retired and moved to Menlo Park, California, not far from Mount Hamilton. Each week he came up to Lick for discussions with all the astronomers and to collaborate in photoelectric research with Kron. They measured in highly accurate detail the light and color variations of the pulsating Cepheid variable stars and were able from their data to reconstruct the temperature, size, and luminosity variation of these objects, so useful for calibrating the scale of the universe. They also compared directly the light and color of the sun and the stars, bridging a brightness ratio of about three trillion, a superb technical feat. Their result is used in all theoretical calculations of luminosities of stars.

Another photoelectric observer, Olin J. Eggen, joined the Lick staff in 1948. Born in Wisconsin, he had received his bachelor's degree at Madison in 1940 and then, after wartime service in the Army, returned and earned his Ph.D. under Stebbins just before he retired. Eggen specialized in precise photoelectric measurements of stars in galactic clusters, which enabled him to plot precise color-magnitude diagrams of these aggregates of stars. These diagrams are somewhat similar to the more traditional Hertzsprung-Russell diagrams of spectral type and magnitude, but because the measured color indices are numerical quantities, the color-magnitude diagrams may be plotted in completely numerical terms. Eggen found that in each cluster there seemed to be not one main sequence as in the earlier Hertzsprung-Russell diagrams, but several very narrow color-magnitude sequences. They promised to reveal new information on stellar evolution, and to make possible much more accurate distance measurements. Eggen's work quickly became widely known and frequently quoted. It was new and controversial. Photoelectric observers at other institutions, particularly Harold L. Johnson at Yerkes and McDonald Observatories, believed their data contradicted Eggen's very narrow sequences. A public controversy followed in which Stebbins, the grand old man of photoelectric photometry, acted as referee. The result was inconclusive at the time, but over the years the concept of multiple, narrow sequences has faded from the literature of galactic cluster research. Eggen left Lick in 1956 to become chief assistant to Astronomer Royal Richard Woolley at the Royal Greenwich Observatory in England.

George H. Herbig, another new staff member in Shane's early years as director, became an outstanding research astronomer. He was a brilliant graduate student and earned his Ph.D. at Lick in 1948 with a thesis on T Tauri stars, the variable stars involved in nebulosity. In the course of this work he discovered the first Herbig-Haro objects, small, very dense nebulae that are intimately connected with star formation. After graduation Herbig was immediately appointed to a junior staff position, but given a year's leave for additional postdoctoral research with Otto Struve, the famous spectroscopist who was director of Yerkes and McDonald Observatories. Herbig returned to Lick in 1949 and made himself one of the world authorities on star formation and the interaction of stars with interstellar matter. He did this work with the Crossley reflector and the 36-inch refractor. Herbig discovered a substantial fraction of the known T Tauri stars, using a slitless spectrograph (of the type now called "grism") that he had built for this program.

It was Herbig who insisted that the 120-inch could not be completed as a bare-bones telescope equipped for research with only a prime focus. During two weeks of each month, when the moon is near full, the sky is too bright for research on very faint stars, which would be done at the prime focus. High-dispersion spectroscopy of bright stars could be done near full moon, but little else. Many discoveries remained to be made in this field, Herbig argued, as the astronomers at Mount Wilson and Palomar were continually proving. It would be foolish not to build a coudé focus and thus to foreclose the possibility of making discoveries at Lick. His arguments won the day and the coudé was added, together with a superb spectrograph that Herbig designed and had built.

Shane presided over the Mount Hamilton community. In 1941 the Birge committee had recommended that the Lick headquarters and staff be moved from Mount Hamilton to Berkeley to share in the scientific contacts as well as the cultural attractions of the campus. Shane had favored this recommendation, and some of the other committee members believed that he would not accept the directorship, after Merrill turned it down, if this policy were not put into effect. It was clearly impossible to carry it out during World War II, but when Shane became director, he stated his intention to make the move. Most of the Lick astronomers favored it, he said, and none opposed it.

However, the very rapid expansion of the Berkeley campus and the postwar inflation did away with this plan. The thousands of new students flocking to Berkeley needed classrooms, the hundreds of new professors needed offices and laboratories, and the many newly created academic departments needed rooms for their headquarters and secretaries. Just keeping up with growth was a problem almost impossible to solve, and the priority for replacement quarters for Lick Observatory, already in spacious existing buildings at Mount Hamilton was obviously low. Shane wanted a good, new building that the Berkeley astronomy department and Lick could share, but by the end of 1948 the university planners decided the needs of other departments were too great to allow this. The astronomy department got a ramshackle collection of wooden temporary buildings left over from the war, and the Lick staff stayed on Mount Hamilton.

There, by 1951, with the engineers, optical shop personnel, and a few construction men working on the 120-inch telescope, together with the regular observatory staff and their families, the population had grown to nearly seventy-five people. There were fifteen family homes, together with the dormitory and mess hall, or diner, for the single men and women. The stage still came up and went down daily, but nearly everyone had a car and could drive down to San Jose in an hour unless the weather was very bad. The one-room school, which had reopened in 1946, typically had eight to ten students through the eighth grade; parents with high-school age children had to make other arrangements. Often they would buy or rent a home in San Jose, where the wife and children would live; the husband would stay on Mount Hamilton during the week, but come down to the valley on the weekend. Sometimes high-school age children went to live with relatives or were sent away to boarding school.

On Mount Hamilton the school building was also a social center for the adults. Wednesday night folk dances were a weekly event in the mountain community, and the graduation ceremonies each June were occasions in which all the students, not just the one or two graduates, had opportunities to show what they had learned by declaiming a speech or poem, or playing a musical piece.

The director's wife, Mary Lea Heger Shane, was herself a Lick Ph.D. who had given up her career when she married and had two children to care for. From a wealthy family, she was an active, outgoing woman with many friends. At Mount Hamilton she became a dedicated hostess to the frequent scientific visitors to the observatory. She and her husband took an active interest in improving the social life of all the inhabitants of the little astronomy village; some of them found it the most wonderful experience of their lifetimes, many enjoyed it, and a few chafed under it.

Shane had not carried out any long-term research program, nor carved out any field as his own in his years at Berkeley. When he became director at Lick, he decided to take over himself the proper-motion survey that Wright had planned. The 20-inch astrograph had been completed, erected, and prepared for operation by Wright before he retired. Shane made the final optical adjustments and began photographing the sky. He and his assistant Carl A. Wirtanen took a total of 1,246 acceptable exposures between 1947 and 1954, covering about 70% of the sky, on 17 × 17 inch photographic plates, each depicting a 6° × 6° area, about the size of the bowl in the Big Dipper. In order to measure the stellar proper motions, a second set of plates had to be taken a few decades later.

Rather than waiting until the second-epoch observations could be started, Shane decided to use the first-epoch plates to count the number of galaxies in the sky as a function of position. Previous knowledge of the distribution of galaxies was based on counts to faint magnitude limits by Hubble and others in only small discrete regions of the sky. They gave a general idea of the distribution, but no possibility of detecting any fine structure in it. It was an ideal program for Shane, as he could fit the routine counting into open times between his administrative duties as director.

Quite early in the program Shane noticed pronounced clustering in the distribution of galaxies, as had been reported earlier from fragmentary data by Harvard astronomer Harlow Shapley. Shane invited Berkeley statisticians Jerzy Neyman and Elizabeth L. Scott to investigate this phenomenon. Their analysis, made jointly with Shane, showed that not only clusters of galaxies but also aggregates of clusters exist in the universe. Shane called these aggregates "clouds," but they are usually known as "superclusters" or "cluster of clusters" today. Thus the by-product of the proper-motion survey turned out to be a very important research program in itself.

Meanwhile, progress on the 120-inch dragged on. The cost had risen to $1.8 million by 1947, and again to $2.2 million in 1952. The first main gear for the telescope was poor, and had to be replaced. There were problems with the figure of the mirror, and eventually Shane brought Hendrix up from Palomar to complete it. There were other problems as well. Baustian had been hired away by the new Kitt Peak National Observatory to be chief engineer for its planned 84-inch reflector. Shane had been willing to let him go, because he had considered that the 120-inch was nearly ready for operation, and thought that Baustian would no longer be needed. But the telescope stubbornly resisted completion. Some of the younger staff members, comparing Shane's early estimate that it could be in operation by 1952 with the reality that it was not in 1957, grumbled that a stronger hand was needed at the helm.

Sproul was due to retire at the end of June 1958. Shane had worked closely with Sproul ever since he had been appointed president in 1930, and decided that he did not want to spend his last few years learning how to get along with a new chief executive. He announced that he would step down as director when Sproul retired. On June 30, 1958, with the 120-inch still not in operation, Shane gave up the directorship of Lick Observatory. His term of thirteen years in office had been second in length only to Campbell's.

14

Santa Cruz

1958–1988

On July 1, 1958, Albert E. Whitford became director of Lick Observatory. He was the first director chosen by the Mount Hamilton astronomers themselves. After C. Donald Shane had decided to step down from the directorship, he informed the Lick faculty well in advance and suggested that they meet without him and nominate a successor. He promised that he would take their recommendation to University of California President Robert G. Sproul, and would support it. They decided on Whitford, Shane recommended him, and Sproul appointed him. Whitford was the first Lick director to come from outside the University of California faculty, except for James E. Keeler, who had previously been a Lick staff member.

Whitford was a product of Wisconsin, born in the little community of Milton, where his father was a professor and then president of Milton College, a small institution steeped in the New England tradition. Whitford earned his bachelor's degree there in 1926, and then went on to the University of Wisconsin as a graduate student in physics. He was excellent in experimental laboratory work, especially in developing sensitive electrical equipment and keeping it operational. Joel Stebbins, director of the Washburn Observatory on the Madison campus, needed a graduate student assistant to find ways to improve his photoelectric photometer, so that he could measure the light of even fainter stars accurately. Stebbins offered Whitford the part-time job, and Whitford was glad to take it, even though it was not in physics, for jobs were scarce during the Great Depression. He soon proved his value to Stebbins by developing an amplifier that could measure much smaller currents than could be reached with the older galvanometers. As he worked with Stebbins, Whitford learned to be a research astronomer.

Stebbins recommended Whitford for a National Research Council postdoctoral fellowship at the California Institute of Technology and Mount Wilson Observatory. His recommendation carried a lot of weight and in 1933 Whitford got the fellowship. He did some spectroscopic research with Ira S. Bowen in the Caltech laboratory, but spent most of his time on photoelectric photometry at Mount Wilson. Whitford met George Ellery Hale at his own private solar observatory in Pasadena, where he was trying unsuccessfully to measure the general magnetic field of the sun. By the end of his two years at Caltech and Mount Wilson, Whitford had decided that he wanted to stay in astronomy.

In 1935 he returned to Madison in a research position at Washburn Observatory, and in 1938 he became an assistant professor on the regular faculty. Whitford worked closely with Stebbins, and their series of joint papers, based on measurements in their six-color photometric system ranging from ultraviolet to infrared, put astronomers' knowledge of interstellar dust and its extinction of the light of distant stars on a sound basis. They made many other important discoveries as well, on Cepheid variables, globular clusters, and galaxies.

During World War II Whitford applied his electronic skills to helping develop radar systems at the Radiation Laboratory on the Massachusetts Institute of Technology campus in Cambridge. He was there from 1941 until 1946, along with a good fraction of the other leading experimental physicists in the United States who were not at Los Alamos, Berkeley, or Chicago working on the atomic bomb. In 1946 he returned to Madison, and when Stebbins retired two years later, Whitford was appointed his successor as director of Washburn Observatory.

Whitford continued Stebbins' tradition of frequent observing trips to the large west coast observatories, especially Mount Wilson. However, Whitford believed that the University of Wisconsin should have a true research observatory near Madison to replace the antiquated 15½-inch refractor that Edward S. Holden had used, and Stebbins also, particularly in his early days at Wisconsin. Whitford's initiative and drive enabled him to convince the University of Wisconsin Alumni Research Foundation to provide funds to build a 36-inch reflector at the Pine Bluff Observatory, in a dark-sky area near Madison. Later Whitford succeeded in getting funds from the National Science Foundation, set up after World War II, for a special grating spectrograph, optimized for photoelectric spectral measurements of nebulae and stars. Pine Bluff was a remarkably advanced midwestern university observatory.

Thus Whitford's whole research career had made him a no-nonsense, hardworking observational astronomer, who had the technical knowledge and drive to produce advanced new instruments and use them to get important new results. And, although he was not a Lick product, he was not really an outsider either for he was the protege of Stebbins, by 1958 the grand old man of the observatory, and he had visited Mount Hamilton often. Gerald E. Kron supported Whitford strongly for the directorship. All the Lick astronomers considered him just the right man to get the 120-inch finished and into operation, and to direct the research they would do with it.

They were right. Whitford took hold of the 120-inch project with a bang. He immersed himself in the project, studying what had been done, and what had gone wrong. He learned all the details, questioning especially Dwight J. "Red" Ludden, the chief engineer who had been appointed when W. W. Baustian left for Kitt Peak in 1957. Whitford was soon making many of the critical engineering decisions himself. He also followed very closely the progress of Nicholas U. Mayall and Stanislavs Vasilevskis, who were testing the mirror in the telescope frequently, as optician Donald O. Hendrix and his assistant Howard Cowan figured it ever closer to the desired exact paraboloidal form. The Hartmann tests that Mayall and Vasilevskis performed each took only a short exposure time, but occasioned days of delay in the figuring process. Each test required transporting the mirror from the optical shop in the basement of the dome to the telescope and then, after the exposure, removing the mirror from the mounting and getting it back safely into the shop, measuring the images on the test plates, and doing all the long numerical calculations to derive the exact shape of the mirror. Vasilevskis worked all night on June 17, 1959, to finish the reduction of the last Hartmann test exposure, and brought the news the next morning to Whitford at the Astronomical Society of the Pacific meeting in San Francisco that the 120-inch mirror was ready for operation.

Whitford's efforts had all paid off. He got the 120-inch reflector completed and doing astronomy. The mirror was finished and was aluminized. The first coudé spectrograms and direct exposures at the prime focus for research programs were taken in the fall of 1959, little over a year after Whitford's arrival at Mount Hamilton, and thirteen years after the project had been approved by the legislature and the governor. The 120-inch was in regular operation by early 1960. At the dedication, Clark Kerr, the new president of the University of California, stated its cost as $2.5 million, more than double the original appropriation. No doubt the costs of the time and energy of the observatory employees that went into it, if they had been included, would have increased this total significantly.

One of the early observations made with the 120-inch was a high-dispersion spectrum that Merle F. Walker of the Lick staff took in collaboration with Andre Lallemand, using the electronographic camera the French astronomer had developed. It was more sensitive than photographic plates, and hence made it possible to get good spectra of fainter stars and galaxies, but it was delicate, fragile, and difficult to keep operational. One of the first discoveries Lallemand and Walker made with it was the rapid rotation of the nucleus of M 31, the Andromeda galaxy, revealed by the Doppler effect on the high-dispersion spectra they were able to obtain. Kron was also working on developing a different type of image tube, also based on the photoelectric effect. Rapid changes in technology were occurring in the 1960s, and Lick astronomers were among the leaders in developing and applying them at the telescope.

George H. Herbig continued his spectroscopic work on star formation and early stellar evolution with the high spectral resolution and superior wavelength coverage of the coudé spectrograph he had built. It permitted astrophysical investigation of the properties of interstellar clouds from which new stars form. He also used the coudé spectrograph to investigate young stars, and the properties they have that have survived from their pre-stellar times. From direct photographs he took with the Crossley and 120-inch telescopes over several decades, Herbig discovered and investigated large proper motions in many Herbig-Haro objects, indicating jet-like motions in these recently formed objects.

Not all the Lick astronomers who had helped to plan the 120-inch stayed long enough to use it. Mayall, who had done so much to make it a reality, was offered the directorship of the new Kitt Peak National Observatory in 1960, and left Mount Hamilton soon after the new telescope went into operation. In 1965 Kron, who twenty years before had accompanied Mayall to the UCLA campus to convince Sproul of Lick's desperate need for a big telescope, departed to become director of the Flagstaff station of the U.S. Naval Observatory. However, both Mayall and Kron were replaced by younger recruits.

George W. Preston was a Lick Ph.D. who spent two years as a "postdoc" (postdoctoral research fellow) at Mount Wilson and Palomar Observatories before returning to Mount Hamilton with a staff appointment. He fitted out the coudé spectrograph on the 120-inch with a magnetic analyzer, modeled after the one Horace W. Babcock had developed at Mount Wilson. With it Preston found new information on stellar magnetism and its connection with stellar rotation. He also made important observational studies of various types of variable stars with the coudé spectrograph.

Thomas D. Kinman, an English astronomer who joined the Lick staff in 1961, was soon using a fast spectrograph at the prime focus of the 120-inch to identify and measure the redshifts of quasi-stellar radio sources ("quasars"). Maarten Schmidt of Caltech, observing with the 200-inch, had first correctly identified the spectral lines in the spectrum of the quasar 3C 273 in 1963. The redshift of all its spectral lines, 43,700 km/sec, was so large that earlier observers had not grasped that they are really the familiar hydrogen lines, seen in an almost completely different spectral region from their normal position in the laboratory and in the spectra of stars and nearby galaxies.

Once Schmidt broke the puzzle, he and other observers with large telescopes were able to obtain spectra of other quasars quickly. They found many with even larger redshifts. The fact that all the quasars were receding from us established that they are distant objects, sharing in the expansion of the universe, which Edwin Hubble had discovered from the redshifts of galaxies. The quasars, as shown by their much larger redshifts, were at even greater distances than the most distant known galaxies, for Hubble, Milton L. Humason, and others had demonstrated that the velocity of expansion increases with distance. Many of the quasars were brighter than galaxies, yet considerably more distant; clearly they must be much more luminous than the galaxies.

Quasars thus provided the opportunity to study the most distant observable reaches of the universe, and Kinman used the fast spectrograph at the prime focus of the 120-inch telescope to do so. E. Margaret Burbidge, who had joined the University of California, San Diego faculty in 1961, also worked on these objects with the 120-inch. Nearly every astronomer had his or her own research program, and a general observatory project, like the old Lick radial-velocity program of Campbell's days, no longer existed.

There was one exception: the proper-motion survey that William H. Wright had planned, and Shane had begun. After he had completed taking the first-epoch plates, Shane had put Vasilevskis in charge of the program. He had first arrived at Mount Hamilton as a postwar immigrant in 1949. Born in Latvia in 1907, Vasilevskis was educated at the University of Riga, where he became first an assistant at the observatory and then a staff member. In 1940 the Red Army marched into Latvia, and Vasilevskis' native land was incorporated into the Soviet Union. Then in 1941 Hilter denounced the Nazi-Soviet pact and invaded Russia; Latvia became part of a new German province called Ostland. In 1944 the Russians fought their way back into Latvia, and Vasilevskis and his family, with thousands of other Latvians, escaped ahead of them into Germany. There Vasilevskis worked at the Leipzig Observatory for the duration of World War II. After the final German defeat, Vasilevskis, his wife, and their two children succeeded in making their way to the relative safety of a displaced persons' camp in the American Zone of Occupation.

They did not want to return to Latvia, already incorporated into the Soviet Union as a "republic." They did not want to remain in Germany. They wanted to emigrate to the United States, but it was not easy after World War II, because so many wanted to get out of Europe. Several church organizations were actively helping displaced persons, cataloguing the skills of would-be immigrants, listing them in the United States, and trying to find jobs and sponsors for them. Vasilevskis taught for a year in Munich, in a temporary refugee university program set up under United Nations' auspices, and registered with the Unitarian Service Commission as a potential immigrant. Augusta Trumpler, the wife of astronomer Robert J. Trumpler, was active in the Unitarian movement in Berkeley, and through it she learned of the Latvian astronomer who wanted to get a job in America. She called Vasilevskis' name to Shane's attention. The Lick director did not know him, but managed to find one of Vasilevskis' astronomical research papers in a volume of the Riga Observatory publications in the Lick Observatory library. It convinced him that the Latvian was a serious astronomer. Shane wanted to help him. All he offered Vasilevskis was a low-paying assistant's job, but the displaced person jumped at it and in 1949 arrived at Mount Hamilton.

Vasilevskis was intelligent, skilled, and very hard working. At first he assisted Hamilton M. Jeffers in taking, measuring, and reducing direct photographs of double stars to determine their orbital properties. Vasilevskis' skill in numerical calculations, his training and knowledge in astrometric research, and his ability to analyze a problem clearly and break it down into its component parts made him an ideal person to take over the task of measuring the Lick proper-motion survey plates. Shane, who had no experience in those lines, handed the program over to him. Vasilevskis' background and practical knowledge enabled him to develop the concepts for a highly automated measuring machine to handle the millions of measurements and reductions necessary for this program. In 1954 Otto Struve, who had come to Berkeley as chairman of its astronomy department, offered Vasilevskis a faculty position there; Shane, by now convinced of how valuable to the observatory the former displaced person was, managed to give him a regular faculty appointment and to keep him at Lick instead.

When Vasilevskis took charge of the proper-motion program in 1954, all of the first-epoch plates had been taken. The second-epoch series of plates could not be started for several years, by which time the effects of the stars' motions would be large enough to be measurable. Vasilevskis immediately recommended that a second 20-inch astrograph lens be designed and constructed to work in yellow light, and mounted in parallel with the existing instrument, which worked in blue light.

This was an idea that Trumpler had put forward twenty years ago, when Wright had first conceived the project. Trumpler had wanted to use the astrograph to measure the magnitudes and colors of stars. Two telescopes that could be pointed at the same star field and exposed simultaneously would be an ideal instrument for this program. Wright had favored having a second telescope built around a lens figured for red light instead of yellow to get a larger "baseline," or difference in color between the two sets of images, but he had started the project with the single blue lens. However, Wright had designed and built the mounting so that a second astrograph could be added later. Shane had ordered the glass for a red lens in 1950, but Vasilevskis knew that yellow would be much better. The blue and yellow combination would give colors of stars directly on the basic photometric system in common use. Even more importantly the exposure times for the blue and yellow lenses could be the same, while a red lens would have required a longer exposure. Vasilevskis drew up the specifications for a yellow-corrected lens. Shane had earlier secured the special glass for a second lens, and James G. Baker of Harvard College Observatory had provided a design. Now the Perkin-Elmer company modified it for yellow light, the National Science Foundation provided funds, and the second lens was made and mounted before the second-epoch exposures began.

In addition to running the proper-motion program, Vasilevskis had other important roles at Mount Hamilton. He had done all the calculations for the testing of the 120-inch mirror from the Hartmann-test exposures that he and Mayall had taken. Vasilevskis started an innovative parallax program with the 36-inch refractor in 1961. It used a new automatic guiding camera, quality control of plate processing, and an automatic measuring machine to determine more accurately the distances of nearby stars. He also began a new astrometric program on the stars in clusters. Vasilevskis had charge of the Lick Observatory library, by now grown so large that an annex had been added to the Main Building to house it. Going back to the books and journals Holden had ordered even before the observatory commenced operations in 1888, it was (and is) one of the great astronomy collections of the world. Cautious and conservative, but highly creative in his thinking, Vasilevskis was a forward-looking scientist and an excellent judge of people. Whitford came to rely on him more and more for counsel and advice on the deepest problems of the observatory.

The problems were not slow in arising. Simultaneously with Whitford's appointment as director, Kerr replaced Sproul as president of the University of California. Kerr, who had a passion for order, launched a study of the organization of the university. It had grown from its original Berkeley location to include other full-fledged campuses at Los Angeles, San Diego, Riverside, and Santa Barbara, and more were

planned. Each campus had its own chancellor, who reported directly to
the president. However, Kerr learned, there were three independent research organizations within the university that also reported to him:
Lick Observatory, the Scripps Institute of Oceanography at La Jolla,
and the Citrus Experiment Station at Riverside. The latter two were the
nuclei around which two of the newer campuses had been built, and
their directors reported through their local chancellors to Kerr. The new
president decided that only chancellors should deal directly with him,
and that each research organization should therefore be attached organizationally to some campus. This meant that Lick would have to
become part of a campus officially. Kerr also believed that all faculty
members of the University of California should do some teaching, and
have at least a minimum interaction with students. It was a proposition
hard to quarrel with, but the Lick astronomers taught no classes and
only interacted with those Berkeley graduate students who came to
Mount Hamilton to observe at the telescopes, either as summer assistants or as Ph.D. thesis candidates.

A possible solution was near at hand, however. The University of
California was growing rapidly, and a new central California campus
was being planned. Though no final decision had been reached, the
university had announced that it would be built in the Almaden Valley,
close to rapidly growing San Jose and not far from Mount Hamilton.
Kerr suggested to Whitford that Lick Observatory be associated with
the proposed Almaden campus, giving it the instant recognition a group
of top scientists and their research tradition would bring. They could
do their teaching there. The Lick astronomers considered this proposal,
but rejected it. They did not want to throw in their future with an
untried and in fact unbuilt new campus in Almaden. The problem remained.

The very success of the 120-inch telescope created new problems for
the Lick observers. Faculty members at Berkeley, Los Angeles, and San
Diego wanted to use the second-largest telescope in the world for their
own research, and were applying for increasing amounts of observing
time with it. More nights for them meant fewer for the Lick astronomers, who began to wonder how large the number of telescope users
would become, as the other campuses grew. Even more serious to the
Lick astronomers was the issue of controlling the telescope and its auxiliary instrumentation. For many years after Lick Observatory went into
operation, it was the only significant research astronomy department in
the university. Later, as Berkeley grew, its astronomy department carefully maintained its subsidiary and complementary role to Lick. Under
Armin O. Leuschner, R. Tracy Crawford, Shane, and Sturla Einarsson
as successive chairmen, the Berkeley department concentrated on teach-

ing and celestial mechanics and never challenged Lick's primacy. However, when Otto Struve, one of the world's great observational astrophysicists, came from Yerkes Observatory to become the chairman on the Berkeley campus in 1950, the situation began to change. Struve did much of his own observing at Mount Wilson, but he started to build up a staff of excellent young research astronomers in Berkeley. When he left to become director of the National Radio Astronomy Observatory in 1959, and Louis G. Henyey succeeded him as chairman in Berkeley, the situation was ripe for change.

In addition, at Los Angeles Daniel M. Popper and Lawrence H. Aller were experienced, well-known research astronomers, and at San Diego, Margaret Burbidge was another skilled observer who used the 120-inch reflector for her research. They could not help but want more voice in the decisions on instruments and observing time being made at Mount Hamilton, which very much affected their own scientific careers. In 1963 President Kerr set up a committee, chaired by Berkeley physicist Robert B. Brode, to advise him on the future of Lick Observatory. The committee members came to Mount Hamilton, asked hard questions, showed little respect for the Lick tradition or research accomplishments, and returned to their campuses to write their report. It was a disaster for Lick.

The Brode committee based its report on teaching. The number of students in American universities was growing by leaps and bounds. There was a great demand for more professors, but the Brode committee found, using statistics from the 1950s, that the University of California was not doing its share in astronomy. The committee recommended that the Berkeley and Los Angeles graduate departments be rapidly built up so that the rate of "production" of Ph.D.s could be doubled. The committee found that the Lick astronomers were taking too much of the observing time, but not doing any of the teaching. Brode's solution was to reduce the Lick staff, by "terminating" (firing) the junior astronomers who did not yet have tenure, and not filling future vacancies that would arise in the course of time "due to retirement" (of tenured Lick astronomers) "or other causes" (deaths). The report emphasized the isolation of the Lick staff on Mount Hamilton, but stated they should be kept there to provide astronomical research facilities for faculty members from the entire University of California system. Administratively, however, they should come under the Berkeley chancellor. The crowning blow was that, throughout the report, the committee referred to new departments and research laboratories, some of them still being built and none of them with proven research records, as if they were equivalent to proud Lick Observatory.

The Lick astronomers were aghast. The whole tenor of the report was an attack on everything they stood for. It ignored their very real research accomplishments and their emphasis on quality, and concentrated instead on counting numbers of students. It would relegate them to a service role, leaving all the teaching and much of the research to the faculty members on the three campuses with graduate astronomy programs. Worst of all, the three bright young Lick astronomers who did not have tenure, Kinman, Walker, and Preston, would all have to go. They were among the best astronomers of their age group in the United States. All three were doing excellent research, but the Brode committee recommended that they not be promoted, and instead be removed from their positions to make room for, in the Lick view, lesser astronomers in the campus departments.

The Lick astronomers regarded the Brode report as an inaccurate, biased document intended to destroy their observatory as a research institution. They reacted angrily, criticizing the report point by point in internal memos. Whitford fought hard against the Brode recommendations, while Shane, who was in his last year before retiring from the faculty, did the same through his informal network of faculty contacts in Berkeley. Nevertheless President Kerr put most of the recommendations into effect, although the junior faculty members were spared, at least for the time being.

Lick Observatory, which had been directly under the university president since 1888, became organizationally a part of the Berkeley campus. Whitford lost his direct channel of communication to the president, and in principle had to go through the Berkeley chancellor with all his requests, recommendations, and budgets; in fact he had to go through the Berkeley dean of the College of Letters and Sciences. This immediately removed much of the decision-making power at Lick Observatory from the hands of the Mount Hamilton astronomers, and transferred it to the Berkeley administrators.

In addition, Kerr set up a new Advisory Committee on Astronomy. It was to be a watchdog "to review the effectiveness of Lick Observatory in meeting its statewide responsibilities" and to advise the president as to the needs "of the several campuses and of the Lick Observatory for appointments in observational astronomy."

The chairman of this new committee was Roger Revelle, director of the Scripps Institute at San Diego, and its members included Emilio Segré, a Berkeley physicist, Robley Williams, a former astronomer turned biophysicist and a severe critic of Lick Observatory, Kron, and Popper. Except for the two astronomers and William H. Rubey, a UCLA geologist, the committee seemed hostile and negative to the Lick faculty. They wanted to continue as a research center, but felt they were being downgraded to a service organization for the rest of the university; they wanted to teach and work with graduate students as in the past, but believed that their opportunities were being usurped by the Berkeley department.

Matters came to a head at an all-university astronomy meeting in San Diego in the spring of 1964. Geoffrey Burbidge, Margaret's husband and a leading theorist on the San Diego faculty, was there. The previous year he had prepared a long statement for the National Academy of Sciences panel, chaired by Whitford, that was studying the future needs of American astronomy. Burbidge's statement had strongly criticized the conservatism of astronomers in general and "the major California observatories" in particular. They did not take theoreticians seriously enough. They stuck "stubbornly" to isolation, and two major observatories, Yerkes Observatory (which Burbidge had recently left) and Lick were still located outside "normal university campuses." They should be dragged "kicking and screaming into the twentieth century" in Burbidge's opinion. Young theoreticians, not old observers, should be appointed as directors of major observatories.

An engaging, effective speaker, Burbidge expressed these thoughts more pointedly at the San Diego meeting. Part of his charm was his gift for unexpected and often extreme phrases, but the Lick astronomers were not charmed in the least by his arguments for wrenching the system they knew so well into a completely new shape. They were appalled when they heard him say at lunch to Kinman, "We've got you on the run, Tom. You can't win. We'll bury you." Of course he did not mean it literally, any more than Nikita Khrushchev had when speaking of the United States a few years before, but it was a memorable phrase and the Lick astronomers did not forget it.

More quietly, Harold F. Weaver, Trumpler's son-in-law and himself a Lick Ph.D. who had been on the Mount Hamilton staff a few years before transferring to Berkeley, made it clear that he considered the mountain observatory's days numbered. The Lick astronomers feared they would have to move to Berkeley in a secondary role, while the Berkeley department would control the graduate students. The San Diego meeting convinced them that they had to do something quickly to break out of the path that they thought would lead to oblivion.

Soon after the meeting Preston came to Whitford and suggested that Lick, instead of moving to Berkeley, become a part of the central California campus that, university planners had now decided, was to be built in Santa Cruz rather than in the Almaden Valley. Whitford and Vasilevskis had independently reached the same conclusion. They encouraged Preston to present his idea to the rest of the Lick staff members. They all agreed that being *the* astronomy department at Santa Cruz would be far preferable to being the second astronomy department at Berkeley. They signed a unanimous letter requesting this transfer, and Whitford, bypassing his official channel of communication through the Berkeley administration, presented it directly to President Kerr. The chief executive was pleased that his original idea would now come true after all, that the new campus would instantly acquire one of the most prestigious research institutions in the country, and that the recriminations between the Lick director, the Brode committee, and the Advisory Committee on Astronomy would now cease. Kerr approved the new plan immediately, and a few days later Vice President Harry P. Wellman, much happier than he had been on previous visits, appeared on Mount Hamilton to work out the details of the transfer. Lick Observatory was going to Santa Cruz.

The new campus, then being planned, was the dream of Dean E. McHenry, longtime UCLA faculty member and administrator and close friend of President Kerr. McHenry had helped to bring about the decision that his new campus would be located in an idyllic redwood forest in the little beach town on Monterey Bay, rather than in the hot Almaden Valley. He stated that he was even more pleased to have Lick Observatory become an integral part of the University of California, Santa Cruz, fulfilling the president's wish. Though the Advisory Committee on Astronomy had to be consulted on the transfer, the result was a foregone conclusion, and on November 20, 1964, the regents officially approved the move.

Shane, though he played no part in arranging the transfer of Lick Observatory to Santa Cruz, helped to expedite it. He had long owned a retreat in Scotts Valley, a community at the edge of the mountains a few miles from Santa Cruz, and when he retired in the summer of 1963, he and his wife had moved there. Several of the Lick astronomers with their families had visited the Shanes over the next year, and had become better acquainted with the attractions of the Santa Cruz area. Shane himself had accepted an appointment as a member of McHenry's campus planning committee, and came to know the chancellor and the other top administrators of the new campus under construction. After the Lick astronomers proposed the move, Shane took the lead in organizing picnics and other informal meetings between his Mount Hamilton and Santa Cruz friends.

In 1964 the campus was under construction. One widely published photograph showed McHenry, a natural showman, sitting at a desk in an open field with beautiful tree-covered mountains in the background behind him. The first, small group of students began classes in September 1965, but they had to live in trailers the first year. Only a few buildings for classes were completed in Cowell College, the first of the colleges that together made up the unique Santa Cruz campus. Vasilevskis, Kinman, and Preston commuted from Mount Hamilton to teach parts of an undergraduate general science course. By the next year a second college, Stevenson, was completed, and in November 1966 the Lick staff began moving to the campus. Initially they were housed in the building then called Central Services, downstairs from the chancellor's office. A tragic accident marred the move; Lick librarian Constance Watson and her daughter were killed in an automobile collision on the winding mountain road between San Jose and Santa Cruz. By the fall of 1967 the Lick astronomers were able to move to their permanent quarters in the unimaginatively named Natural Sciences II building, across the road from the relocated Lick shops. The faculty, secretaries, and business and shop personnel all moved to Santa Cruz, leaving only the telescope operations staff and maintenance personnel at Mount Hamilton.

Among the new Lick faculty members were E. Joseph Wampler, who joined the staff in 1965, and Joseph S. Miller, in 1967. Wampler was a Yerkes Ph.D. who had been a postdoc at Lick for two years before getting the faculty appointment. He was a gifted instrumentalist, and developed a photoelectric scanning spectrometer for measuring accurately the radiation in individual spectral lines or at specific wavelengths in the spectra of stars and nebulae. He recruited Lloyd B. Robinson, trained as an experimental physicist at the University of British Columbia, to work with him on this and other instrumental projects. This first "Wampler scanner" gave Lick Observatory one of the most advanced astronomical instruments of the late 1960s, and Wampler himself, Miller, a Wisconsin Ph.D., and other staff members from throughout the University of California used it to advantage on stars, nebulae, and galaxies.

At Santa Cruz, Lick Observatory retained its role as a research organization, and also served as the nucleus of a new department of astronomy. In the UCSC terminology it was the Board of Studies in Astronomy and Astrophysics, and Preston was named its first chairman. The Lick astronomers could teach the observationally oriented graduate courses, but they needed to add some theoreticians for the more mathematical subjects. They hoped to get several additional faculty positions in which to hire these theoreticians, but McHenry, who was building Santa Cruz as an innovative, undergraduate-teaching oriented campus, would not assign them any. He felt he needed all the faculty positions that the state and the university would give him for teachers of undergraduate courses. Therefore Lick hired Peter H. Bodenheimer, who had earned his Ph.D. as a theoretician at Berkeley in 1966, to a faculty post. That fall the first graduate student, Diane M. Pyper, transferred to Santa Cruz from Berkeley, where she had completed all her course work. She did her thesis on magnetic stars under Preston's direction, and earned the first UCSC Ph.D. in astronomy in 1968. Graduate astronomy classes at Santa Cruz began in 1967, with students Jack Baldwin, Robert R. Zappala, Larry R. Cathey, William R. Alschuler, and Robert Milton.

McHenry's refusal to assign any new faculty positions to the Board of Studies in Astronomy and Astrophysics was only the first sign of trouble between his administration and the Lick faculty. Their aims were completely different—innovative undergraduate education versus creative hard-science research—and clashes were bound to occur. This natural tension was greatly exacerbated by Francis H. Clauser, the first vice-chancellor of the Division of Science and Engineering at UCSC. His scientific background, aims, and ideas were so contradictory to Whitford's and the Lick staff's that the clashes were severe, and soon led to the departure of the main actors from the center of the stage.

Clauser had done his undergraduate and graduate work in aeronautical engineering at Caltech, where he made a brilliant record. Later, while he was vice-chancellor at UCSC, his alma mater listed him as one of the twenty-five outstanding graduates in its history. After earning his Ph.D. in 1937, Clauser became a designer and research aerodynamicist for the Douglas Aircraft Corporation in southern California. After World War II he went to Johns Hopkins University as chairman of its Aeronautics Department from 1947 until 1960, and continued four more years there as a senior professor. Then in 1965 he came to Santa Cruz as the prospective head of the engineering program McHenry intended to start. All the science boards of studies, as well as Lick Observatory, reported to the chancellor through Clauser. With his background in aerospace, the vice-chancellor naturally wanted to launch UCSC into the world of space astronomy, which was expanding rapidly in the 1960s. He said that he respected Lick Observatory's achievements, but all his actions indicated to the astronomers that he would have much preferred them to forget their traditional ground-based 120-inch reflector on Mount Hamilton, and jump into the space race. His dream was to bring an early version of the Space Telescope project to the Santa Cruz campus. He hoped to bypass all the proposals and peer reviews that by 1965 had become normal in NSF funding of astronomical research, and follow instead the aerospace industry's tradition of dealing directly with the heads of NASA. Whitford, who always leaned over backwards to follow the rules, knew that Clauser's approach would not work, but could not convince the vice-chancellor. All the Lick astronomers wanted to get on with using the new 120-inch telescope for research, rather than becoming involved in space projects whose scientific payoff would be far in the future. They tried to ignore Clauser and boycotted his appeals for their cooperation in his plans.

He became increasingly preemptory in his dealings with them. They were used to being supported, not to being critically reviewed. Soon Whitford and Preston as chairman of the board were involved in almost perpetual struggles with Clauser.

Finally, in February 1968, Preston, fed up with the general tension with the vice-chancellor, and disturbed in particular by the delay of a faculty promotion he had recommended, resigned his faculty position and took a job at Mount Wilson and Palomar Observatories. They had been after him for several years, and he decided that he had had enough of the struggle with Clauser. Shortly thereafter Whitford, worn down by what he considered McHenry's lack of support for Lick Observatory, resigned the directorship, effective June 30, 1968. He had had enough fights also. By then, projections showed that the national need for engineers was decreasing rapidly, and the California Coordinating Council on Higher Education had decided that there should not be a new engineering school at UCSC. Clauser stepped down as vice-chancellor also on June 30, 1968, and departed for Caltech a year later.

Robert P. Kraft was appointed acting director to replace Whitford. Kraft, a Lick Ph.D. of 1955, had afterward been a postdoc at Mount Wilson and Palomar, and then briefly a faculty member successively at Indiana University and Yerkes Observatory. He had returned to the Mount Wilson and Palomar staff in 1960, and had accepted a faculty position at Lick in 1967, after the move to Santa Cruz. He was an expert in stellar spectroscopy, making particularly important contributions to the study of Cepheid variable stars and of old novae. Kraft was willing to serve as acting director for a year, but did not want to accept the post on a permanent basis.

A university-wide search committee for a new director was set up. Its members included Charles H. Townes, inventor of the maser and a Berkeley professor of physics, Geoffrey Burbidge, and George O. Abell of UCLA, as well as Herbig of the Lick staff. They recommended various candidates, some of whom were approached but declined, and in the end Kraft, who was doing an excellent job, had to agree to remain as acting director for a second year. In the fall of 1969 the University of California finally offered the position to Donald E. Osterbrock of the University of Wisconsin. Whitford had tried to get him to come to Lick as a visitor the previous year, just after he had resigned the directorship himself, but Osterbrock was already committed to a sabbatical year at University College London, in England. McHenry had visited him there in the spring of 1969, and had told him that he wanted to bring him to Santa Cruz as director of Lick.

Osterbrock, however, declined the appointment in the fall of 1969 and then again in December of that year, after it had been offered to him a second time. He wanted to go to Lick, but he wanted even more to stay in Madison. Kraft went on sabbatical leave to the Joint Institute for Laboratory Astrophysics at the University of Colorado in the summer of 1970, and Herbig became acting director for a year. The position was offered to one or more other astronomers, who declined it, and Kraft became acting director again on September 1, 1971. In 1972 Osterbrock finally accepted the post. He was a Yerkes Ph.D., who had specialized in nebular photometry and spectroscopy at Caltech, Mount Wilson and Palomar Observatories and at the University of Wisconsin. On July 1, 1973, he became the ninth director of Lick Observatory.

Under Kraft and Osterbrock, relations between the observatory, the UCSC administration, and the astronomy departments on the other campuses gradually improved. Among the many astronomers who used the 120-inch, some of the best known were Margaret Burbidge of UCSD, Hyron Spinrad and Ivan King of UCB, and Lawrence H. Aller and Daniel M. Popper of UCLA.

In 1967 Preston and Kraft had conceived the plan of applying to the NSF for a departmental improvement grant, to make it possible to hire more theoreticians on the Board of Studies in Astronomy and Astrophysics faculty. McHenry supported their proposal strongly, and UCSC got the grant. Under its terms one new faculty member could be hired each year for five years. The NSF paid the first several years' salary for each; UCSC agreed to provide the funds to keep the positions filled thereafter. Under this grant Jeffrey Scargle, John Faulkner, William G. Mathews, and others later joined the faculty. The grant, which was stretched out to cover six years, provided additional funds for secretarial help and for graduate student support. Kraft became chairman after Preston left, and later the post was rotated every few years among the various senior theorists. Within a few years Santa Cruz became a flourishing graduate school of astronomy, turning out many bright young Ph.D.s in astronomy who took research positions all over the United States. David M. Rank, a Michigan Ph.D. who spent three years as a postdoc at Berkeley working with Townes, brought expertise in infrared astronomy to the Lick faculty in 1971. Then Sandra M. Faber, a Harvard Ph.D., joined the Lick faculty in 1972, Steven S. Vogt, from the University of Texas in 1978, and Burton F. Jones, a Yerkes Ph.D., in 1979.

In 1971 Wampler and Robinson put into operation on the 120-inch reflector the image tube-image dissector spectral scanner they had developed at Santa Cruz. It was a fast, electronic, digital system that enabled Lick observers to make accurate spectrophotometric measurements of faint stars, galaxies, and nebulae. It was mounted at the Cassegrain focus of the telescope, ideally suited for a large, remotely operated spectrograph. The new "Wampler-Robinson scanner" quickly became one of the most-used instruments on the telescope, and made Lick Observatory the envy of the astronomical world. In 1974 a greatly improved spectrograph designed by Miller especially for the image tube-image dissector system went into operation, making it even more flexible and useful. Results on abundances of the elements in stars and nebulae, on Seyfert galaxies and quasars, on variable stars and interstellar matter flowed out of Mount Hamilton. Observers from all the campuses participated in the research.

In 1981 after more than eight years as director, Osterbrock stepped down, and Kraft, now willing to accept the post, followed him. Under his directorship the University of California finally succeeded in its search for funds for another telescope, larger than the 120-inch (which was renamed the Shane reflector in 1978) and larger even than the 200-inch Hale telescope at Palomar.

As the campus astronomy departments grew in the 1960s, all the University of California astronomers had felt the need of another large telescope to carry the load of the increasing number of observational research workers in the system. Ever since W. W. Campbell's days, a large southern-hemisphere telescope had been a dream at Lick Observatory, and Whitford, soon after becoming director, began planning in that direction. It was an all-university effort, with Lick playing the lead role. In 1963, when this project became active, the principal continental European nations were already planning a European Southern Observatory, and the Mount Wilson and Palomar astronomers, working through the Carnegie Institution of Washington, had high hopes of erecting their own southern-hemisphere observatory. Both were planned for Chile. The Lick astronomers therefore thought in terms of a site in Australia, which has a different weather pattern from South America. Campbell had considered an Australian site for the Lick southern station in 1901, but had rejected it because of the long sea journey necessary to reach it; by 1963 air travel had reduced the distance to a minor problem. In 1964 Whitford sent C. Robert O'Dell, of the Berkeley faculty, to Australia for several months on a detailed site survey. By then it was clear that the University of California could not provide the funds necessary to build a large telescope, and that they would have to be sought outside the state budget. Whitford hoped for a collaboration with the Australians, some of the money to come from their government and some from the NSF or NASA.

In 1966, however, the Ford Foundation, which had been on the point of providing money for the Carnegie telescope in Chile, combined instead at the last moment with the NSF to fund a U.S. National Observatory in Chile, which eventually became Cerro Tololo Interamerican

Observatory. Then Australia agreed with England to build a joint An-
glo-Australian telescope at Mount Stromlo, Australia, putting an end
to the possible California collaboration. This eliminated any hope of
funds for the projected California southern-hemisphere telescope, and
the Lick astronomers' thoughts increasingly turned closer to home.
Walker had long advocated seeking a new observatory site in the Santa
Lucia mountains between the Pacific Ocean and the Salinas Valley,
south of Monterey. They were similar in weather pattern and stable
atmosphere to Mount Hamilton, but far from any potential source of
light pollution. This had become a great problem at Lick, as the population in San Jose and the Santa Clara Valley grew, and the "miracle
of electricity" increasingly dominated American life. After a preliminary
site survey, Walker selected Junipero Serra Peak, at an elevation of
5,700 feet in the Los Padres National Forest, as the prime site and
testing continued there for several years. Nearly a century before, when
it was still known as Mount Santa Lucia, George Davidson, James
Lick's astronomical adviser, had observed a total solar eclipse from
there.

Whitford, Kraft, Herbig, and Osterbrock all devoted considerable
time and energy to the Junipero Serra project. They tried unsuccessfully
to raise money from various foundations, or through combinations with
other universities. Fairly large sums from the University of California
went into an environmentally sound plan to build the observatory
within the highly protected wilderness area. However, none of the directors could ever secure the much larger amount of money necessary
to actually build the telescope. It was seen as a "100-inch class" telescope at first, which gradually grew in the planning to a 150-inch, but
the basic problem was that by the 1970s there were many telescopes
of that size already in existence. No person or agency seemed willing
to finance one more of them.

However, Wampler, and independently Jerry Nelson, an astrophysicist at the Lawrence Berkeley Laboratory, became convinced that the
University of California should build a telescope significantly larger than
the 200-inch. To construct such a large telescope would mean adopting
some completely new principle, for a conventional design would be far
too expensive. Wampler's idea was to make the mirror much thinner
than in a conventional telescope, and hence the whole mounting much
lighter and less expensive. Nelson, in contrast, wanted to abandon the
concept of a single mirror, and instead cut the weight and thus the costs
by assembling a mosaic of hexagonal glass segments, which together
would form the primary mirror. Both plans, the "monolith" and the
"mosaic," were imaginative in taking advantage of all the advances in
technology and design capability that had occurred since the Palomar
telescope was planned in the 1930s.

Kraft took Nelson and Wampler to see President David S. Saxon of the University of California in the fall of 1977, while Osterbrock was on a short sabbatical leave at the University of Minnesota. Saxon, a former theoretical physicist, was especially taken with Nelson's mosaic approach, and authorized an in-depth study of it and of Wampler's plan. The project was an all-university one, with the Lick director taking a lead role. Everyone favored making a 400-inch, or 10-meter telescope. Many of the technical facets of the two competing designs were investigated at Berkeley and at Santa Cruz. Osterbrock appointed a "graybeards" committee, consisting of senior astronomers from Santa Cruz, Berkeley, Los Angeles, and San Diego to make the decision as to which path to follow. At a meeting in the Berkeley Faculty Club on November 27, 1979, chaired by Osterbrock, they voted to go with Nelson's segmented design.

The university fund raisers swung into action, but success was slow in coming. At last, however, on December 15, 1983, Marion Hoffman agreed to give $36 million to the University of California to build the "largest telescope in the world," the planned 10-meter telescope. She was the widow of a very successful car importer, and she first had heard of the telescope project from her brother, who in turn had learned of it from W. J. Shiloh Unruh, amateur Lick Observatory historian. Hoffman announced her gift at a meeting with David P. Gardner, who had succeeded Saxon as president, but tragically she died the next day. She knew that she was suffering from inoperable cancer, but no one had expected the end to come so soon.

Unfortunately her magnificent gift was not enough to build the telescope, for its cost was estimated at $90 million. It proved impossible to raise the additional amount from University of California donors, and eventually President Gardner sought partners from other academic institutions. The California Institute of Technology was the first to come forward. One of the Caltech trustees, Howard B. Keck, chairman and president of the W. M. Keck Foundation, offered to provide not just the remainder of the needed funds, but nearly the entire amount. The Hoffman Foundation then withdrew from the project, and now the W. M. Keck Observatory with its 10-meter telescope is being erected on Mauna Kea, Hawaii, by the California Association for Research in Astronomy, consisting of the University of California and Caltech. When it is completed the observatory will belong to Caltech, but the University of California has committed itself to provide the operating funds. The two universities will use the 10-meter telescope jointly and share equally in the observing time.

Mauna Kea was chosen for the 10-meter telescope because its 13,600-foot elevation, stable atmosphere, relatively mild climate, low latitude, and freedom from light pollution make it the best astronomical site in the United States. Thus the main University of California observatory will no longer be on Mount Hamilton, after the Keck Observatory is completed. But undoubtedly the Lick Observatory staff will play a large part in operating the 10-meter telescope and make important discoveries about the nature of the universe, just as their predecessors did for a century on Mount Hamilton. And the 120-inch Shane telescope, on what is still a very good site except for light pollution, will continue working at the frontiers of astronomical research for many years to come.

15

One Hundred Years On

1888–1988

Lick Observatory's outstanding importance is that it was the first big-science research observatory in America. Its development and evolution over the past century have mirrored the progress of astronomy. Planned from the beginning for use by a large staff of investigators for their individual research programs, the institution was built around the 36-inch telescope, a large, expensive piece of capital equipment financed by James Lick.

The Lick refractor was the first large telescope erected at a site chosen for its astronomical advantages, rather than for convenience in the builder's backyard or on a university campus. For this reason, Lick Observatory was built on a mountain, and it turned out to be an excellent choice, one followed by the most successful observatories erected from 1888 until the present day. Lick, the man who made the decision, and George Davidson, his first adviser, understood some of the advantages of this site. It had the transparent air of the highest available mountain within Lick's range of interest, and it was in California, the land of clear skies. Other advantages they did not understand, but were lucky to have nevertheless. Mount Hamilton, like other excellent observatory sites, is a sharp, knife-edged mountain that causes little perturbation or turbulence in the laminar airflow that comes in over the cool ocean current to its west. Only in very recent years have such sites, in both the northern and southern hemispheres, been surpassed by Hawaii's Mauna Kea, a high mountain that penetrates up into the stable air over the middle of a large ocean.

When the 36-inch refractor went into operation, "superior to and more powerful than any telescope yet made," as required by Lick's deed of trust, it established Lick Observatory as the premier observatory in the world. Within less than ten years the Yerkes 40-inch refractor surpassed the Lick refractor in aperture, but Lick retained its preeminence in research results, for the Wisconsin site was much poorer than Mount Hamilton, both in the amount of clear weather and in atmospheric steadiness (good seeing). Also, as Lick's first director, Edward S. Holden, had foreseen, with the Crossley 36-inch reflector augmenting the refractor after 1898, Mount Hamilton could again legitimately claim the superior telescope power.

In 1908 when the Mount Wilson 60-inch reflector went into operation in southern California, it pushed Lick Observatory into second place. Ten years later Mount Wilson's 100-inch reflector more than doubled its advantage. Nevertheless, even though several other large reflectors subsequently went into operation—at the Dominion Astrophysical Observatory in Canada, at Perkins Observatory in Ohio, and at McDonald Observatory in Texas—the skill and dedication of the Lick staff, the auxiliary instruments they designed and built, and the advantages of the Mount Hamilton site kept Lick Observatory in second place for many years.

After World War II, when the Lick staff had long recognized that its major telescope, the Crossley, was outdated, the Palomar 200-inch became the largest and most powerful telescope in the world. Little more than a decade later the 120-inch raised Lick again into second place, and its astronomers began one of their most productive eras. By the end of the 1970s the 120-inch had been surpassed by several 150-inch telescopes, and by 1988 in size it was only one of the top fifteen telescopes in operation in the world. In addition, the Mount Hamilton site has deteriorated over the years, due to the growth of Silicon Valley and its accelerated light pollution. Without the cooperation of San Jose and other nearby municipalities in replacing mercury streetlights with low-pressure sodium vapor lamps, Lick might have been effectively blinded by now. In spite of these setbacks, in research productivity Lick is still one of the most important observatories in the world.

As telescopes and technology improved over the years, astronomers applied them to important new areas of research, and Lick led the way in several. The Crossley reflector that Holden brought to Mount Hamilton and that James E. Keeler put into use was the first "large" reflecting telescope used by professional astronomers at a first-class site. With it Keeler opened the eyes of his generation to spiral "nebulae" as important constituents of the universe. W. W. Campbell, impressed by the Crossley, built and got into operation the 36-inch reflector of the Lick southern station in Chile as a cheap, powerful light collector for spectroscopy. All subsequent larger successful telescopes have followed these research paths. They had been anticipated in the minds of George Ellery Hale and George W. Ritchey at Yerkes Observatory, but Keeler and Campbell were the first to get large reflectors into regular use for research.

Keeler's direct photographs taken with the Crossley were surpassed pictorially by Ritchey's when the Yerkes astronomer put his 24-inch reflector, with its superior mounting and better optics, into operation in 1901. But the Crossley still reigned as the supreme research reflecting telescope until Ritchey completed the 60-inch in 1908. Even then, Heber D. Curtis, working with the smaller Crossley, led the way in pushing forth the understanding of spiral "nebulae" as galaxies, giant island universes composed of stars. Only when Edwin Hubble went to work with the 100-inch reflector after World War I, and actually proved that galaxies contain stars, and then discovered the expansion of the universe, was Lick surpassed in this field.

Curtis was but one of many first-rate researchers who worked at Lick over the years. Lick's first director, Holden, was a dismal failure in his dealings with human beings, but an outstanding success in picking the original staff. Three of them, S. W. Burnham, E. E. Barnard, and Keeler were among the best research astronomers of their time. Barnard's great visual discovery, the tiny fifth satellite of Jupiter, caught the imagination of the public because it came three centuries after Galileo's discovery of the first four. It made Barnard and Lick Observatory famous. Charles D. Perrine subsequently discovered two more photographically, and Seth B. Nicholson still another one at Lick.

Burnham and after him Robert G. Aitken were the outstanding dou‑ble-star observers in America; in this field Lick was unsurpassed for many years. Hamilton M. Jeffers continued this work, but by his time the main results were already known, and his research had less impact. John M. Schaeberle and Richard H. Tucker did meridian-circle work, positional measurements of stars, but this is a type of research not dependent on a big telescope and hence was not unique to Lick. It was never a high-priority program at Mount Hamilton, and little came of it.

Keeler's astrophysical investigations were highly creative, carefully planned and carried out, and led to important new ideas on the physical nature of the universe. He was on the brink of even more important discoveries when he died.

His successor, Campbell, changed the direction of Lick Observatory research from pioneering astrophysical investigation to the steady accumulation of radial velocity data, but the methods he used and force he possessed made it an outstandingly successful program. He made himself one of the dominant figures in American astronomy for over a quarter of a century. Campbell's assistants in the radial velocity program, William H. Wright, Joseph H. Moore, and Curtis, all went on to become highly successful astronomers, each in his own way. Wright followed Keeler's path, accurately measuring the wavelengths of many nebular emission lines, and thus providing the raw material for later identifications by Caltech's Ira S. Bowen. Curtis led the way in understanding the spirals. Moore followed Campbell's example, and became the radial velocity expert and cataloguer of the next decade.

Campbell also organized and led many eclipse expeditions, which took him all over the world, but he was always too busy to do much with the spectra and direct photographs of the corona and chromosphere obtained at the expense of so much time and effort. Reduced finally by Donald H. Menzel, they opened the era of detailed astrophysical study of the outer layers of the sun. But this program would have been far more productive if Campbell had found someone like Menzel twenty years earlier, so that the observational material could have been analyzed soon after each expedition. In that way the lessons learned from one eclipse could have been used to improve the observational methods, especially of photometric calibration, before the next. Campbell's recognition (or acceptance) of the necessity for a precise test of Einstein's predictions of the gravitational deflection of light matched Lick eclipse expertise with an important physical problem. After failures at Brovary and Goldendale, Robert J. Trumpler's reduction of the Wallal eclipse plates conclusively confirmed the general theory of relativity.

Trumpler's pioneering photometric studies of star clusters led to his recognition of the extinction of light in interstellar space, one of the most important new results in twentieth century astronomy. Yet despite having Menzel on the staff and Bowen as a Morrison Fellow, astrophysics never really got started at Lick Observatory in the pre-World War II years. The senior staff members were too firmly set in their ways, and the telescopes were too outdated. The leaders of the observatory, all Campbell's former colleagues and assistants, were too old and too inflexible. And they were too proud of themselves and of their heritage; too distrustful of foreigners, and even of Americans with names that they considered "foreign."

Yet Wright had the vision of the Lick proper-motion program, based on a reference system defined by the distant galaxies. And he had the drive to get the telescope that made the program a reality. As time ran out on him, he wanted to sacrifice the idea of a new big telescope to divert all the resources of the observatory to the proper-motion program. But he lost out to the arguments of youth and to a university president who realized that he had to spend money to build a large telescope for the outstanding research observatory he wanted.

Lick's other great contribution to the past century of astronomy has been its involvement with graduate students, the astronomers for to-morrow. From Keeler and Armin O. Leuschner down to C. Donald Shane, graduate training in research was always one of Lick Observatory's strongest points. The combination of courses on the Berkeley campus and observational work with the telescopes on Mount Hamilton was unbeatable. Probably Leuschner emphasized celestial mechanics too long, and Wright and Moore were too slow in promoting astrophysics, but for many years a large number of the best American observational astronomers came from the University of California. The program was originally conceived by Keeler and Leuschner to provide new staff members for Lick and Berkeley, and it did. Later, students left for Mount Wilson and Palomar, too, and for Michigan and Wisconsin and Carleton and San Mateo, as it turned out.

The final pages in the history of Lick Observatory are yet to be written. After World War II, the promise of the 120-inch reflector revitalized Lick, even years before the new telescope was completed. Lick astronomers took the lead in developing new instruments and new observing techniques, particularly in photoelectric photometry, which had been pioneered by former Lick student Joel Stebbins. When the 120-inch did go into operation, second in size only to the 200-inch telescope, new results followed in every field from nearby white dwarf stars to distant galaxies. Posterity can classify and judge the individual scientists and research programs, but the Lick Observatory of this era will surely not be found wanting in scientific productivity.

It has been one of the great observatories in America for a century; never truly the greatest after its first twenty-five years, but always close to the top. Astronomy and America are the richer because Lick Observatory has existed so long and has done so much.

References

ARCHIVAL SOURCES

Bancroft Library, University of California, Berkeley
    Raymond T. Birge Papers
    George Davidson Papers
    Richard S. Floyd Papers
    Phoebe Apperson Hearst Papers
    Eugene W. Hilgard Papers
    Armin O. Leuschner Papers
    University of California Archives
George Ellery Hale Papers
    Microfilm Edition, California Institute of Technology, Pasadena
Lick Observatory Plate Vault, Mount Hamilton
    Observing books of all astronomers and programs
Manuscript Division, Library of Congress, Washington, D.C.
    Asaph Hall Papers
    Simon Newcomb Papers
Mary Lea Shane Archives of Lick Observatory
    University Library, University of California, Santa Cruz
Mount Wilson Observatory Archives, Pasadena
    Walter S. Adams Papers
Townley Family Papers, Palo Alto
    Sidney D. Townley Papers
University of Wisconsin Archives, Memorial Library, Madison
    College of Letters and Science, Department of Astronomy
Yerkes Observatory Archives, Williams Bay, Wisconsin
    Director's Papers

The primary sources for the history of Lick Observatory are the letters written to and from the officers of the Lick Trust, going back to 1873, and to and from the astronomers of the observatory, primarily the director, going back to 1888. They are for the most part preserved in the Mary Lea Shane Archives of the Lick Observatory, located in the McHenry Library on the campus of the

University of California, Santa Cruz. This archives was conceived, organized, and directed for many years by the late Mary Shane, who was a Lick Ph.D. in astronomy who gave up her scientific career when she became a housewife and mother. More than any other person, she was responsible for preserving the documentary history of Lick Observatory.

Letters concerning Lick Observatory and its astronomers exist in many other archives as well, and were consulted in the course of preparing this book. Rather than clutter the text with references, we describe briefly in the following notes the main sources for each chapter. References to specific letters are contained in many of the books we cite. In addition, we used many contemporary news-paper clippings from the time, preserved in a series of scrapbooks going back to the first director, Edward S. Holden. And, of course, the scientific papers, the fruit of the astronomers' research, provided an important historical source. These papers are published primarily in the *Lick Observatory Publications* (*L. O. Pub.*), *Lick Observatory Contributions* (*L. O. Cont.*), *The Astrophysical Journal* (*Ap. J.*), and *Publications of the Astronomical Society of the Pacific* (*P.A.S.P.*), and in other scientific journals as well.

Books used as sources for many chapters in the present work include:

Gingerich, Owen W., ed. *Astrophysics and Twentieth Century Astronomy to 1950*. (*The General History of Astronomy, 4*). Cambridge: Cambridge University Press, 1983.

Osterbrock, Donald E. *James E. Keeler, Pioneer American Astrophysicist: and the Early Development of American Astrophysics*. Cambridge: Cambridge University Press, 1984.

Wright, Helen. *James Lick's Monument: The Saga of Captain Richard Floyd and the Building of the Lick Observatory*. Cambridge: Cambridge University Press, 1987.

Important historical articles written by participants in the history of Lick Observatory include:

Aitken, R. G. "The Lick Observatory, Forty Years After," *P.A.S.P.*, 40, 151, 1928.

Holden, Edward S. "The Lick Observatory," *Sidereal Messenger*, 7, 49, 1888.

Moore, J. H. "Fifty Years of Research at the Lick Observatory," *P.A.S.P.*, 50, 189, 1938.

Shane, C. Donald. "Lick Observatory: The First 75 Years," *P.A.S.P.*, 76, 77, 1964.

Wright, W. H. "The Founding of Lick Observatory," *P.A.S.P.*, 50, 143, 1938.

1. HIS TOMBSTONE IS A TELESCOPE, 1848–1876

The story of James Lick is told in:

Anonymous. "The Life of James Lick," *Quarterly of the Society of California Pioneers*, 1, 14, 1924.

Lakey, Edward C. "California's Miser Philanthropist: A Biography of James Lick," M.A. Thesis, University of California, 1949.

Lick, Rosemary. *The Generous Miser: The Story of James Lick of California.* Palo Alto: Ward Ritchie Press, 1967.

Worrilow, William H. *James Lick (1796–1876): Pioneer and Adventurer: His Role in California History.* New York: Newcomen Society, 1949.

Lakey's thesis is by far the best of these and contains the most complete references, but many areas of Lick's early life are still and will probably always remain subjects for conjecture. The anonymous account that appeared in the Society of California Pioneers' journal is thought by many to have been written by Henry Mathews, long-time secretary to the Lick Trust.

The life of George Davidson is recounted in:

Campbell, W. W. "The Astronomical Activities of Professor George Davidson," *P.A.S.P.,* 26, 28, 1914.

Lewis, Oscar. *George Davidson, Pioneer West Coast Scientist.* Berkeley and Los Angeles: University of California Press, 1954.

George Madeira's claims to have first shown James Lick a telescope in 1860 and then again in 1873, and to have suggested to him that he found an observatory, are recounted in his letters to Edward S. Holden, especially the one dated July 14, 1887, located in the Shane Archives, but there is no known contemporary evidence to support them.

2. THE GREATEST ASTRONOMER OF HIS TIME, 1874–1878

Simon Newcomb wrote an entertaining autobiography that can be used as a first reference for researching his life, but a more recent paper is far more reliable:

Newcomb, Simon. *The Reminiscences of an Astronomer.* Boston & New York: Houghton, Mifflin & Co., 1903.

Norberg, Arthur L. "Simon Newcomb's Early Astronomical Career," *Isis,* 69, 209, 1978.

Much of Newcomb's correspondence on behalf of the Lick Trust is preserved in the Shane Archives and in his own papers in the Library of Congress in Washington, D.C.

Information on the founding and history of the Naval Observatory can be found in:

Herman, Jan K. "The Establishment of the U.S. Naval Observatory," *Vistas in Astronomy,* 28, 391, 1985.

Herman, Jan K. "A Hilltop in Foggy Bottom," pamphlet produced by the Naval Medical Command, Department of the Navy, 1984.

Some insight into Joseph Henry, Secretary of the Smithsonian Institution, and an account of his involvement with James Lick can be found in:

Jones, Bessie Zaban. *Lighthouse of the Skies.* Washington: Smithsonian Institution, 1965.

3. A SON OF THE CONFEDERACY, 1876–1884

4. THE TELESCOPE COMES TO THE MOUNTAIN, 1884– 1888

The most complete account of Richard S. Floyd's life and involvement with Lick Observatory is contained in Helen Wright's book mentioned above. Additional information on the construction of the observatory can be found in:

Chriss, Michael. "The Stars Move West: The Founding of the Lick Observatory," *Mercury*, 2, 10, 1973.

Holden, Edward S. *L.O. Pub.*, 1, 1887.

Neubauer, F. J. "A Short History of Lick Observatory, Part 1," *Popular Astronomy*, 58, 201, 1950.

Information on Sherburne W. Burnham can be found in:

Burnham, S. W. "A General Catalogue of 1299 Double Stars Discovered from 1871 to 1899 by S.W. Burnham," *Publications Yerkes Observatory*, 1, vii, 1900.

Barnard, E. E. "Sherburne Wesley Burnham," *Popular Astronomy*, 29, 309, 1921.

Frost, E. B. "Sherburne Wesley Burnham," *Ap. J.*, 54, 1, 1921.

Macpherson, Hector, Jr. *Astronomers of To-Day and Their Work*. London: Gall & Inglis, 1905.

Burnham's report on his testing of Mount Hamilton's observing conditions is contained in:

Burnham, S. W. "Report of Mr. S. W. Burnham," *L.O. Pub.*, 1, 13, 1887.

The history of the 12-inch refractor is given in:

Warner, Deborah Jean. *Alvan Clark & Sons, artists in optics*. Washington: Smithsonian Institution Press, 1968.

A full account of Warner and Swasey's involvement is given in:

Pershey, Edward Jay. "The Early Telescope Work of Warner and Swasey." Ph.D. Dissertation, 1982. Case Western Reserve University.

An account of James E. Keeler's discovery of a new gap in Saturn's rings during the first night of use of the 36-inch refractor is given in:

Osterbrock, Donald E., and Cruikshank, Dale P. "Keeler's Gap in Saturn's A Ring," *Sky and Telescope 64*, 123, 1982.

5. INTO THE OCEAN OF SCIENCE, 1888–1895

6. THE GREAT I AM, 1888–1897

A report on Edward S. Holden's years as director and on the first Lick Observatory staff is given in:

Chriss, Michael. "Of Stars and Men: Lick Observatory's First Decade of Operation," *Mercury*, 2, 3, 1973.

Neubauer, F. J. "A Short History of the Lick Observatory, Part 2," *Popular Astronomy*, 58, 318, 1950.

Some insight into Holden himself comes from:

Osterbrock, Donald E. "Edward S. Holden—The Founder of the A.S.P.," *Mercury, 7,* 106, 1978.
Osterbrock, Donald E. "The Rise and Fall of Edward S. Holden," *Journal for the History of Astronomy, 15,* 81, 151, 1984.
Winlock, William C. "Sketch of Professor Edward S. Holden," *The Popular Science Monthly,* 114, 1886.

Reports on astronomer E. E. Barnard and his scientific activities are contained in:

Barnard, E. E. *L.O. Pub., 11,* 9, 1913.
Burnham, S. W. "Edward Emerson Barnard," *Popular Astronomy, 1,* 193, 341, 441, 1894.
Cruikshank, Dale P. "Barnard's Satellite of Jupiter," *Sky and Telescope, 64,* 220, 1982.
Griffin, Roger F. "Barnard and his Observations of Io," *Sky and Telescope, 64,* 428, 1982.
Macpherson, Hector, Jr., *Astronomers of To-Day and their Work.* London: Gall & Inglis, 1905.

7. ADONAIS, 1897–1900

A full account of James E. Keeler's life and short career is given in the book by Donald E. Osterbrock mentioned above.

Information on Samuel P. Langley is provided by:

Beardsley, Wallace R. "Samuel Pierpont Langley: Early Conflict between Teaching and Research at the Western University of Pennsylvania," *Western Pennsylvania Historical Magazine, 64,* 345, 1981.
Holden, Edward S. "Sketch of Professor S. P. Langley," *Popular Science Monthly, 27,* 401, 1885.

The history and arrival at Lick Observatory of the Crossley reflector is told in:

Stone, Remington P. S. "The Crossley Reflector: A Centennial Review Part 1," *Sky and Telescope, 61,* 307, 1979.

"Adonais, an Elegy on the Death of John Keats" was written by Percy Bysshe Shelley in 1819. It mourns the untimely death of Adonais, whose "fate and fame shall be an echo and a light unto eternity."

8. THE CREATIVE SCIENTIST WHO BECAME A FACTORY MANAGER, 1900–1923

The life of W. W. Campbell is told in:

Wright, W. H. "Biographical Memoir of W. W. Campbell 1862–1938," *Biographical Memoirs, National Academy of Sciences, 25,* 35, 1949.
Aitken, Robert G. "William Wallace Campbell," *P.A.S.P., 50,* 204, 1938.

Heber D. Curtis' career is recounted in:

McMath, Robert R. "Heber Dorst Curtis 1872–1942," *P.A.S.P.*, *54*, 69, 1942.

Aitken, Robert G. "Heber Dorst Curtis 1872–1942," *Biographical Memoirs, National Academy of Sciences, 22*, 275, 1943.

Stebbins, Joel. "Heber Curtis and the Michigan Telescope," *Publications of the Observatory of the University of Michigan, 10*, 1, 1951.

Two excellent historical papers on Curtis' concepts of spiral "nebulae" as galaxies are:

Hoskin, M. A. "Ritchey, Curtis and the Discovery of Novae in Spiral Nebulae," *Journal for the History of Astronomy, 7*, 45, 1976.

Hoskin, M. A. "The 'Great Debate': What Really Happened," *Journal for the History of Astronomy, 7*, 169, 1976.

Also useful are the early chapters of the book:

Smith, Robert W. *The Expanding Universe: Astronomy's "Great Debate,"* *1900–1931*. Cambridge: Cambridge University Press, 1982.

In addition there are thousands of letters to and from Campbell in the Mary Lea Shane Archives of Lick Observatory, including a relatively small file of personal material, and very many letters from Curtis. Their research can be traced in their many published papers. Perrine's life is less well documented, and was reconstructed largely from letters about him (plus a few of his own) in the Shane Archives, together with his fewer (and shorter) research papers. Day-to-day life on Mount Hamilton is described in the weekly letters of Richard H. Tucker to his mother and sister, in a reminiscence provided by his daughter Molly Tucker Beer, in a recorded interview of Mary McDonald Wilson by John R. Gustafson and W. J. Shiloh Unruh, and in:

Campbell, Kenneth. "Life on Mount Hamilton 1899–1923," Interviewed and Edited by Elizabeth S. Calciano, Santa Cruz, 1971.

9. IN THE SHADOW OF THE MOON, 1889–1930

On James C. Watson and Vulcan, see:

Comstock, George C. "Memoir of James Craig Watson 1838–1880," *Biographical Memoirs, National Academy of Sciences, 3*, 43, 1895.

On Edward S. Holden's eclipse expeditions before he came to Lick Observatory:

Osterbrock, Donald E. "The Rise and Fall of Edward S. Holden: Part 1," *Journal for the History of Astronomy, 15*, 81, 1984.

The Lick Observatory eclipse expeditions are described in popular articles and news items in the *P.A.S.P.*, except for the Bartlett Springs and Cayenne eclipses, which occurred before this journal was founded. They are both reported at length in the *L.O. Cont.* Scientific papers giving the immediate results of all of the eclipse expeditions from 1898 onward are published in the *Ap. J.* and the *L. O. Pub.* The long paper by Donald H. Menzel giving the results of his analysis of the chromosphere, based on the eclipse data, is in *L. O. Pub.*

There are many letters concerned with these eclipse expeditions in the Shane

Archives, especially reports and instructions exchanged between the parties in the field and the Lick headquarters on Mount Hamilton. An especially rich source for all the expeditions in which Elizabeth B. Campbell participated are her letters to her mother, children and friends, her eclipse diaries, and her book-length manuscript "In the Shadow of the Moon," all of which are held in the archives.

In addition, the following historical articles were useful in the preparation of this chapter:

Crelinsten, Jeffrey. "William Wallace Campbell and the Einstein Problem: An Observational Astronomer Confronts the Theory of Relativity," *Historical Studies in the Physical Sciences, 14,* 1, 1984.

Earman, John, and Glymour, Clark. "Relativity and Eclipses: The British Eclipse Expeditions and Their Predecessors," *Historical Studies in the Physical Sciences, 11,* 49, 1980.

Eddy, John A. "The Schaeberle 40-ft Eclipse Camera of Lick Observatory," *Journal for the History of Astronomy, 2,* 1, 1971.

Menzel, Donald H. "Quantitative Chemical Analysis and the Structure of the Solar Atmosphere," *Annals of the New York Academy of Sciences, 198,* 235, 1972.

Osterbrock, Donald E. "Lick Observatory Solar Eclipse Expeditions," *Astronomy Quarterly, 3,* 67, 1980.

Zirker, Jack B. "Testing Einstein's General Relativity During Eclipses of the Sun," *Mercury, 14,* 98, 1985.

10. LEUSCHNER'S LEGACY, 1888–1931

Biographical articles on some of the Lick Observatory graduate students whose stories are told in this chapter are:

Makemson, Maud W. "Russell Tracy Crawford 1876–1958." *P.A.S.P., 71,* 503, 1959.

Moore, J. H. "Ralph Hamilton Curtiss 1880–1929." *P.A.S.P., 42,* 37, 1930.

Anonymous. "Edward A. Fath." *P.A.S.P., 71,* 258, 1959.

Alter, Dinsmore. "Armin Otto Leuschner 1868–1953." *P.A.S.P., 65,* 269, 1953.

Herget, Paul. "Armin Otto Leuschner." *Biographical Memoirs, National Academy of Sciences, 49,* 129, 1978.

Joy, Alfred H. "Paul Willard Merrill 1887–1961." *P.A.S.P.* Wilson; O.C. "Paul Willard Merrill." *Biographical Memoirs, National Academy of Sciences, 37,* 237, 1964. *74,* 41, 1962.

Herget, Paul. "Seth Barnes Nicholson." *Biographical Memoirs, National Academy of Sciences, 42,* 201, 1971.

Nicholson, Seth B. "Frank Elmore Ross 1874–1960." *P.A.S.P., 73,* 182, 1961.

Morgan, W. W. "Frank Elmore Ross." *Biographical Memoirs, National Academy of Sciences, 39,* 391, 1967.

Wilson, Ralph E. "Roscoe Frank Sanford 1883–1958." *P.A.S.P., 70,* 360, 1958.

Whitford, Albert E. "Joel Stebbins." *Biographical Memoirs, National Academy of Sciences, 49,* 293, 1978.

Aitken, Robert G. "Sidney Dean Townley 1867–1946." *P.A.S.P.*, *58*, 193, 1946.
Further material on the lives and careers of Stebbins and Townley, particularly their student days, is included in:
Osterbrock, Donald E. "Further Links in the California-Wisconsin Connection." *Transactions Wisconsin Academy of Sciences, Arts and Letters*, *69*, 153, 1981.
In addition, there are many letters to, from, and about each of these students in the Shane Archives. Many of these letters were written to or by Leuschner. Townley's student days are especially well documented in his diary and the letters between himself and his future wife, Frances Wright. They are in the Townley Family Papers, Palo Alto, California. The autobiographical notes of C. Donald Shane provided additional material for this chapter.

11. DOUBLE-STAR OBSERVER, 1923–1935

Biographical articles on some of the scientists treated in this chapter are:
Jeffers, Hamilton M. "Robert Grant Aitken 1894–1951," *P.A.S.P.*, *64*, 5, 1952.
Menzel, Donald M. "Quantitative Chemical Analysis and the Structure of the Solar Atmosphere," *Annals of the New York Academy of Science*, *198*, 235, 1972.
Osterbrock, Donald E. "Nicholas T. Bobrovnikoff and the Scientific Study of Comet Halley 1910," *Mercury*, *15*, 46, 1986.
Shapley Harlow. "Henry Norris Russell 1877–1957," *Biographical Memoirs, National Academy of Sciences*, *32*, 354, 1958.
Weaver, Harold, and Weaver, Paul. "Robert Julius Trumpler 1886–1956," *P.A.S.P.*, *69*, 304, 1957.
In addition there are many letters to and from all these astronomers in the Shane Archives, as well as in the University of California Archives in Berkeley. Their published scientific papers are the basis for the descriptions of the research they did.
Winston Churchill's visit to Lick Observatory is described in:
Gilbert, Martin. *Winston S. Churchill*, Volume 5, *The Prophet of Truth 1922–1939*. Boston: Houghton Mifflin Co., 1977.
and also in the *San Jose Mercury Herald*, *San Francisco Chronicle*, and *San Francisco Examiner* of September 12, 1929.
Several letters from and interviews with Louis Berman, as well as an interview with Ernest H. Cherrington, Jr., by Donald E. Osterbrock provided additional material for this chapter.

12. THE LAST OF THE FIRST, 1935–1945

Biographical articles on some of the Lick astronomers described in the chapter include:

Merrill, Paul W. "William Hammond Wright 1871–1959." *P.A.S.P.*, 71, 305, 1959.

Shane, C. D. "William Hammond Wright 1871–1959." *Biographical Memoirs, National Academy of Sciences*, 50, 377, 1979.

Wright, W. H. "Arthur Bambridge Wyse 1909–1942." *Ap. J.*, 97, 89, 1942.

Mayall, Nicholas U. [Autobiographical chapter] in "Then There Was Light, Autobiography of a University, Berkeley: 1868–1968," edited by Irving Stone, p. 107. Garden City, N.Y.: Doubleday & Company, 1970.

Aitken, Robert G. "Joseph Haines Moore: 1878–1949. A Tribute." *P.A.S.P.*, 61, 125, 1949.

The material on Lawrence H. Aller is based upon personal letters, interviews, and the edited transcript of an interview he had with David H. De Vorkin, August 1979, on file at the American Institute of Physics, New York. Some of the material on William H. Wright and Joseph H. Moore was drawn from the autobiographical notes of C. Donald Shane in the Shane Archives of Lick Observatory.

The airplane crash at Lick Observatory is very fully reported in the *San Jose Mercury, San Francisco Chronicle*, and other Bay Area newspapers of May 18, 1939 and the following several days. A brief account of it was given by:

Wright, W. H. "Airplane Crash at the Lick Observatory," *P.A.S.P.*, 57, 172, 1939.

Correspondence not only from the Shane Archives but also from the University of California Archives was most useful for all parts of this chapter.

13. SECOND LARGEST TELESCOPE IN THE WORLD, 1945–1958

The life and scientific career of C. Donald Shane are recounted in:

Vasilevskis, S., and Osterbrock, D. E. "Charles Donald Shane," *Biographical Memoirs, National Academy of Sciences*, in press.

The life of George Ellery Hale and the early history of Yerkes, Mount Wilson, and Palomar Observatories are given in the excellent scientific biography:

Wright, Helen. *Explorer of the Universe. A Biography of George Ellery Hale.* New York: E. P. Dutton & Co., 1966.

Additional material is from the voluminous correspondence between Hale and successive directors of Lick Observatory in the Shane Archives, and the microfilm edition of Hale's papers.

This chapter is also based on letters and other documents in the Shane Archives, but at the time of writing only a few were available for the period after 1945, as most of the director's files after that year were still in storage. They were supplemented by C. Donald Shane's autobiographical notes, and transcripts of two long interviews he gave Helen Wright and Elizabeth S. Calciano, which are also in the Shane Archives. We also used notes from conversations with the late C. Donald Shane, and from interviews of Albert E. Whitford, Stanislavs Vasilevskis, and George H. Herbig by Donald E. Osterbrock, letters of Nicholas U. Mayall and Gerald E. Kron to Osterbrock, and many newspaper clippings filed in the Shane Archives.

In addition, material from the Lick Observatory files in the University of California Archives in Berkeley was very helpful. It was supplemented by material from the Raymond T. Birge Papers in the Bancroft Library, as well as correspondence between Birge and Otto Struve in the Yerkes Observatory Archives, relating to the Lick Observatory directorship in the years 1940–1945.

14. SANTA CRUZ, 1958–1988

An excellent autobiographical account is:
Whitford, A. E. "A Half-Century of Astronomy," *Annual Review of Astronomy and Astrophysics*, 24, 1, 1986.
At the time of writing of this book, there was almost no material for these years in the Shane Archives, except for a very few official letters and reports, and the extensive collection of letters from C. Donald Shane to Nicholas U. Mayall from 1960 until the early 1980s. These were supplemented by the fuller, but still incomplete material in the Lick Observatory files in the University of California Archives. At the time of writing they were available only for the years through 1966. In addition newspaper clippings and news notes in the *P.A.S.P.* were useful. Interviews of William P. Bidelman, Geoffrey Burbidge, George H. Herbig, Dean E. McHenry, C. Robert O'Dell, Stanislavs Vasilevskis, Harold F. Weaver, and Albert E. Whitford by Donald E. Osterbrock provided further information, as did his own notes, letters, calendars, journal, and memories of events in which he was involved from 1968 until the present.

Index

Yerkes, Charles T., 95, 215, 232
Yerkes Observatory, 126, 182, 243, 257; Bobrovnikoff at, 211; dedication of, 118–119; establishment of, 95–96, 215, 232–233; Kuiper at, 203; Ross

at, 223–224
Young, Charles A., 185

Zengeler, Emil, 122–123

3C 273, 251

Designer:	U.C. Press Staff
Compositor:	Auto-Graphics, Inc.
Text:	10/12 Sabon
Display:	Sabon
Printer:	Edwards Bros., Inc.
Binder:	Edwards Bros., Inc.